A CASUAL REVOLUTION

A CASUAL REVOLUTION

Reinventing Video Games and Their Players

Jesper Juul

The MIT Press
Cambridge, Massachusetts
London, England

For information about special quantity discounts, please email special_sales@mitpress.mit.edu

This book was set in Scala Serif and Scala Sans on 3B2 by Asco Typesetters, Hong Kong.
Printed and bound in the United States of America.

Library of Congress Cataloging-in-Publication Data

Juul, Jesper, 1970–
A casual revolution : reinventing video games and their players / Jesper Juul.
    p.   cm.
Includes bibliographical references and index.
ISBN 978-0-262-01337-6 (hardcover : alk. paper) 1. Video games—Psychological aspects. 2. Video gamers—Psychology. I. Title.
GV1469.34.P79J88   2010
794.8—dc22   2009009091

10 9 8 7 6 5 4 3 2 1

Contents

Acknowledgments

First, I must thank colleagues and students for the knowledge, helpfulness, and discussions that made this book infinitely better than it would have been otherwise.

This book was made possible by the support of the Singapore-MIT GAMBIT Game Lab; Comparative Media Studies; and the School of Humanities, Arts, & Social Sciences at the Massachusetts Institute of Technology.

Thanks to Doug Sery and MIT Press for supporting this project. Thanks to Philip Tan for creating an inspiring environment at GAMBIT. Thanks to Brennan Young, Jaroslav Švelch, Eitan Glinert, Doris Rusch, Jonas Heide Smith, Nick Fortugno, Annakaisa Kultima, and Eric Zimmerman for commenting on the text in various stages. Special thanks to Susana Tosca, Miguel Sicart, Clara Fernández-Vara, and Jason Begy. Thanks to the developers interviewed: David Amor, Sean Baptiste, Daniel Bernstein, Jacques Exertier, Nick Fortugno, Frank Lantz, Garrett Link, Dave Rohrl, Warren Spector, Margaret Wallace, and Eric Zimmerman. Thanks to Gamezebo for collaborating on a survey of their users, and thanks to the players who participated.

Thanks also to John Sharp and Mads Rydahl for 2-D illustrations. Thanks to Geoffrey Long for package photographs.

Thanks to Nanna Debois Buhl for an illustration and everything else.

The arguments in the book have been presented in various embryonic stages at the Serious Games Summit of the Game Developers Conference in San José, Indiana University Bloomington, Georgia Institute of Technology, University of California, San Diego, Stanford University, Vienna University of Technology, Teacher's College of Columbia University, the

Clash of Realities conference in Köln, Massachusetts Institute of Technology, the ACE 2007 conference in Salzburg, the Game Philosophy Conference in Potsdam, and the [player] Conference in Copenhagen.

Three chapters were previously published in earlier versions:

- Chapter 2 was published as "What Is the Casual in Casual Games?" in *Proceedings of the [player] Conference* (Copenhagen: IT University of Copenhagen, 2008), 168–196.
- Chapter 4 was published as "Swap Adjacent Gems to Make Sets of Three: A History of Matching Tile Games" in *Artifact Journal* 2 (2007), 205–216.
- Chapter 7 was published as "Without a Goal" in Tanya Krzywinska and Barry Atkins, eds., *Videogame/Player/Text* (Manchester: Manchester University Press, 2007), 191–203.

Additional material can be found on the website of this book: http://www.jesperjuul.net/casualrevolution.

1 A Casual Revolution

Spending the winter of 2006–07 in New York City, I was beginning to lose count of the times I had heard the same story: somebody had taken their new Nintendo Wii video game system home to parents, grandparents, partner, none of whom had *ever expressed any interest whatsoever* in video games, and these non-players of video games had been enthralled by the physical activity of the simple sports games, had enjoyed themselves, and had even asked that the video game be brought along for the next gathering. What was going on?

When I dug a little deeper, it turned out that many of the people I thought were not playing video games in fact had a few games stored away on their hard drives. These were not shooting games or big adventure games, but smaller games—matching tile games, games about running restaurants, games about finding hidden objects in pictures, and, of course, Solitaire. These players did not fit any stereotype of the adolescent male video game player. In fact, they often did not think of themselves as playing video games (even though they clearly were).

The office and holiday parties of that year were also dominated by a new musical game with plastic guitars, and it dawned on me that this was not about video games becoming *cool*, but about video games becoming *normal*. Normal because these new games were not asking players to readjust their busy schedules. Normal because one did not have to spend hours to get anywhere in a game. Normal because the games fit the social contexts in which people were already spending their time, normal because these new games could fulfill the role of a board game, or any party game.

This looked like a seismic change, but when I asked people why they had not played video games before, another pattern emerged. Many of these people I'd thought were playing video games for the first time would on closer questioning happily admit to having played much earlier video games like *Pac-Man* and *Tetris*, and to having enjoyed them immensely. Hence the bigger picture was not just that video games were finding a new audience, but also that video games were *reconnecting* with an audience that had been lost. Why? The answer: the first video games had been made for a general audience because there was no separate audience of game experts at the time. Between the arcade games of the early 1980s and today, video games have matured as a medium, developed a large set of conventions, grown a specialized audience of fans...and alienated many players.

The casual revolution in the title of this book is a breakthrough moment in the history of video games. This is the moment in which the simplicity of early video games is being rediscovered, while new flexible designs are letting video games fit into the lives of players. Video games are being reinvented, and so is our image of those who play the games. This is the moment when we realize that everybody can be a video game player.

The Pull of Games

As an avid video game player, I have experienced much of the first thirty years of video game history first hand, and it has been disconcerting to see great games ignored by many potential players. Given that video games are as wonderful as they are, why wouldn't you play them? The best way to answer this may be to consider what it feels like to enjoy video games. This experience, of being a *gamer*, can be described as the simple feeling of a *pull*, of looking at a game and wanting to play it. Consider the jigsaw puzzle shown in figure 1.1. In all likelihood you know how you would complete it. You can imagine the satisfaction of moving the final piece, of finishing the puzzle. The jigsaw begs you to complete it.

Or look at the video game shown in figure 1.2. If you have ever played *Pac-Man*,[1] you know your mission is to eat the dots and avoid the ghosts, and from a brief glance at the screen, you may already have planned where you want to go next in the game.

This is the pull of video games, and indeed, of nondigital games too. You can see what you need to do in the game, you can see, more or less, how to do it, and you *want* do to it. In music, or in stories, we experience

Figure 1.1
Complete the puzzle (image ©kowalanka–Fotolia.com)

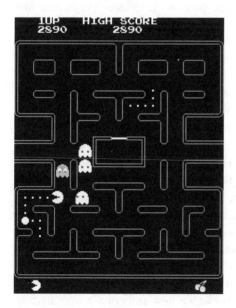

Figure 1.2
Pac-Man (Namco 1980)

Figure 1.3
WarCraft III (Blizzard 2002)

a similar type of pull: When Frank Sinatra sings *"I did it my—"* we want him to end the melody on *"way."* There is a pull toward the final note of the song, the *tonic* in musical terms. A story's pull makes us want to know what happens, how the characters deal with the situation, or who committed the crime. These things pull us in. Video games are like stories, like music, like singing a song: you want to finish the song on the final note. You must play this game. You *must.*

Why must you? The video game's pull is a subjective experience that depends on what games you have played, your personal tastes, and whether you are willing to give the game the time it asks for. For example, who can resist being moved by the invitation of the game shown in figure 1.3? A real-time strategy game is waiting to be played.

Actually, many people do not feel any pull whatsoever toward playing this game. Perhaps *you* do not. The illustrated game, *WarCraft III,*[2] is not universally loved. While it is fairly certain that you know what a jig-saw puzzle asks of you, and there is a high chance that you know what to do with the game of *Pac-Man*, a modern game like *StarCraft* is divisive. Not everybody feels the pull: not everybody knows what to do, not everybody wants to pick up the game and start playing.

This I have always found perplexing, so this book is the result of my journey toward understanding that mystery of why somebody would choose not to play video games, and why a new audience is *now* starting to play video games. I am going to tell stories of the players and developers who are part of the casual revolution, and I will show how changing game designs are reaching new players.

By now I do understand why some would not feel that pull. I understand the frustration of not knowing which buttons to push, of being unfamiliar with the conventions on the screen, of being reluctant to invest hours, days, and weeks into playing this game, of being indifferent to the fiction of the game, of having a stupid machine tell you that you have failed, of *being unable to fit a game into your life.*

A Casual Game for Every Occasion

There is a new wave of video games that seem to solve the problem of the missing pull; games that are easy to learn to play, fit well with a large number of players and work in many different situations. I will refer to these new games using the common industry term *casual games*. In this book I am focusing on the two liveliest trends in the casual revolution:

• The first trend is games with *mimetic interfaces*. In such games the physical activity that the player performs mimics the game activity on the screen. Mimetic interface games include those for Nintendo Wii (see figure 1.4), where, for example, playing a tennis video game involves moving your arm as in actual tennis. Other examples include music games such as *Dance Dance Revolution,*[3] *Guitar Hero*[4] (figure 1.5), and *Rock Band.*[5]

• The second trend is known as *downloadable casual games*, which are purchased online, can be played in short time bursts, and generally do not require an intimate knowledge of video game history in order to play. Figure 1.6 shows the downloadable casual game *Cake Mania 3.*[6]

When I refer to these trends I use the term *video games* to describe all digital games, including arcade games and games played on computers, consoles, and cell phones. Video games reach players through a number of different distribution channels. Whereas mimetic interface games are generally console games sold in stores, downloadable casual games are sold on popular websites. While the increasing reach of video games can also be witnessed in the popularity of small, free, browser-based games like *Desktop Tower Defense,*[7] the focus here is on the commercially more successful mimetic interface and downloadable casual games.

In the short history of video games, casual games are something of a revolution—a cultural reinvention of what a video game can be, a

Figure 1.4
Nintendo Wii players (Saul Loeb/AFP/Getty Images)

Figure 1.5
Guitar Hero II player (AP/Wide World Photos/D. J. Peters)

Figure 1.6
Cake Mania 3 (Sandlot Games 2008)

reimagining of *who* can be a video game player. A manager from the video game publisher Electronic Arts describes the challenge of creating games for a new audience as a *rewiring* of the company: "I was surprised by how wired we were to a particular target audience of 18–34-year-old guys. It was a challenge to change the rule book of designing games for fraternity brothers."[8]

The rise of casual games also changes the conditions for creating games targeted at non-casual players. A game designer describes it as "harder and harder to find people willing to fund games that only go after that narrow hardcore audience."[9] In other words, the rise of casual games has industry-wide implications and changes the conditions for game developers, pushing developers to make games for a broader audience. The rise of casual games influences the development of other video games as well.

Does this go beyond a few high-profile games? Are video games really reaching out to a broad audience? The answer is yes. The Entertainment

Software Association reports that 65 percent of U.S. households play video games today, and that the average age of a game player is 35 years.[10] In the United Kingdom, a BBC report says that 59 percent of 6- to 65-year olds play one form of video game or another.[11] These numbers are growing,[12] and are likely to continue to grow: a recent report shows that a staggering 97 percent of the 12–17 age group in the United States play one form of video games or another.[13] Not that every single person in the world is playing video games just yet, but we can imagine a future where that would be the case. The simple truth is that in the United States and many Asian and Western countries, there are now more video game players than non-video game players. *To play video games has become the norm; to not play video games has become the exception.*

Games and Players

Simple casual games are more popular than complex hardcore games.[14] Casual games apparently reach new players, and the new players they reach are often called *casual players*. But what is casual? The concepts of casual players and casual games became popular around the year 2000 as contrasts to more traditional video games, now called *hardcore* games, and the hardcore players who play them. Casual players are usually described as entirely different creatures from hardcore players:

There is an identifiable *stereotype of a hardcore player* who has a preference for science fiction, zombies, and fantasy fictions, has played a large number of video games, will invest large amounts of time and resources toward playing video games, and enjoys difficult games.

The *stereotype of a casual player* is the inverted image of the hardcore player: this player has a preference for positive and pleasant fictions, has played few video games, is willing to commit little time and few resources toward playing video games, and dislikes difficult games.

To what extent do these stereotypes map to actual players? Surprisingly, when studies were carried out, they showed that more than a third of the players of downloadable casual games played nine two-hour game sessions a week.[15] Effectively, it seemed that casual players were not playing in casual ways at all. This raised a question: do casual players even exist? Looking at the games commonly described as casual yields a clue in that these games allow us to have a meaningful play experience within a short time frame, but do not prevent us from spending more time on a game.

More traditional hardcore design, on the other hand, requires a large time commitment in order to have a meaningful experience, but does *not* allow a meaningful experience with a shorter commitment. It then follows that the distinction between hardcore and casual should not be treated as an either/or question or even as a sliding scale, but rather as a number of parameters that can change over time because players change over time. The stereotypical casual player gradually acquires a larger amount of knowledge of video game conventions, effectively making the player more like a stereotypical hardcore player in terms of game knowledge. The stereotypical hardcore player, conversely, may find that he or she has less time to play video games due to growing responsibilities, jobs, and children, and so that player's willingness to make time commitments diminishes over time, effectively pushing the player toward more casual playing habits.

To discuss casual games and casual players, it therefore becomes important to avoid the temptation to choose *between* them. There are two possible starting points:

1. Start with *games*: to examine the design of casual games.
2. Start with *players*: to examine how and why casual players play video games.

On the one hand, given that some players play casual games in what we could hardcore ways, it could be tempting to conclude that a game can be played in any way players desire, and that game design as such can therefore be ignored. On the other hand, many players tell stories of how casual games are the only video games they will play, so it would be futile to ignore the games. In my opinion, the idea of having to choose between players and games is a dead end. Instead I take as my starting point the way games and players *interact with, define, and presuppose each other*. A player is someone who interacts with a game, and a game is something that interacts with a player; players choose or modify a game because they desire the experience they believe the game can give them. Seeing games and players as mutually defined makes it clearer why some people do, or do not, play video games.

Though they were never quite true, conventional prejudices say that all video game players are boys and young men. A common (and also imprecise) assumption about casual games is that they are only played by women over the age of 35.[16] In early descriptions, the women playing casual games were assumed to play only occasionally and with little time investment. Seeing that this is often not the case, the usefulness of taking gender or age as a starting point for discussing players becomes uncertain.[17] Furthermore, the interviews with game players conducted for this book show that changing life circumstances are major influences on the interviewees' playing habits: reaching adolescence, having children, getting a job, having the children move away from home, and retiring all led to major changes in game-playing habits. The question of how games fit into people's lives is therefore the primary angle in this book.

Many video games ask for a lot in order to be played, so it is not surprising that some people do not play video games. Video games ask for much more than other art forms. They ask for more time and they more concretely require the player to understand the conventions on which they build. A game may or may not fit into a player's life. A game may require hardware the player does not have or does not wish to own, it may build on conventions that the player does not know, require skills the player does not have; it may be too easy for a player or too hard, it may not be in the *taste* of the player. Different games ask different things from players, and different players are not equally willing to give a game what it asks.

Games as well as players can be flexible or inflexible: where a casual game is flexible toward different types of players and uses, a hardcore game makes inflexible and unconditional demands on the skill and commitment of a player. Conversely, where a casual player is inflexible toward doing what a game requires, a hardcore player is flexible toward making whatever commitment a game may demand. This explains the seeming paradox of the casual players making non-casual time commitments: a casual game is sufficiently flexible to be played with a hardcore time commitment, but a hardcore game is too inflexible to be played with a casual time commitment.

Changing Games, Changing Players

Game audiences and game designs co-evolve. The audience learns a new set of conventions, and the next game design can be based on the assumption that the audience knows those conventions, while risking alienating those who do not know them. Where video game developers have often been criticized for making games "for themselves," casual game developers are encouraged to make games for an audience they are not necessarily part of. Designing for players with little video game experience places conflicting pressures on game developers between innovating enough to provide an experience the player recognizes as worthwhile, and at the same time building on only well-known conventions in order to reach a broad audience. This does not render innovation impossible, but means that innovation often has to be based on the import of culturally well-known activities—such as tennis or guitar playing.

It would be wrong to say that casual games were inevitable, but in hindsight it is clear that many things paved the way for them. The first decades in the history of video games saw video games mature as a medium and develop an elaborate set of conventions that has made them inaccessible to potential players unwilling to commit the time to learn these conventions. Strategy and action games, for example, use a number of interface conventions to communicate the events in the game, making this information easily accessible to those who know the conventions, but presenting a barrier to players new to them. When video games developed a new expressive and creative language of their own, they also shut out people who did not know that language.[18] That is the big story of the history of video games and the rise of casual games. For casual players, there are many smaller stories to tell.

There is, for example, the story of the person who never played video games, and now with casual games finds video games that he or she enjoys. A casual game player in her fifties told me she had played board games and card games all of her life, but had only started playing casual games, and video games at all, after being introduced to *Zuma* by a friend:

My 75-year-old friend introduced me to *Zuma* and *Collapse*, the predecessor to *Zuma*. It was after I had handed in my thesis, so my brain was completely offline. Then she invited me over for dinner and told me she had something interesting to show me. She also had a computer Mahjong game that was very beautiful and exciting, I really liked that. Later I have begun to buy them myself, because they are not that expensive.[19]

Then there is the story of the player who avidly played console and arcade games as child, stopped playing video games as they became more complicated, and returned to them via casual games:

When I was a kid, I played *Pong*.... Fast-forward about 20 years. Now I'm married and have children.... They, of course, have video game systems. To me, these systems look like Mission Control for NASA, so I never play with them. I can't. There are too many buttons.

I can play Wii games. The controller is instinctive to use. In fact, the WiiMote is actually easier to operate than the remote control for my television. *WiiBowl* requires two buttons: A and B. That's totally my speed.... With the advent of a gaming system that doesn't require an advanced degree to operate, I have been able to rediscover the joy I found in those early video games I played as a kid. I've found a way to bond with my own children over something that interests them, and when [my] extended family gets together, we have multigenerational play. It's been a great way for my kids, my spouse and I, and my parents to find common ground.[20]

There is also the story of the player who grew up with video games and now has a job and children, making it difficult to integrate traditional video games into his or her life, creating a demand for titles that require less time to play. One self-termed "ex-hardcore-now-parent" player describes the situation like this:

That pretty much sums up my situation these days. Snatched moments are far more child friendly than hour-long *Mass Effect* sessions. That doesn't mean I don't like sneaking off upstairs to have a bit of [Xbox] 360 time but I can have a game of *Mario Kart* or *Smash Bros* and it's literally five minutes while my daughter entertains herself. Maybe that is the market that the Wii has tapped into. Not the non-gamer; more the ex-hardcore-now-parent gamer.[21]

My own story intersects the big story of casual games, and is also a story of changing life circumstances: I have a life-long love for video games and I have spent much time trying to convince friends and family to play them. Casual games work so much better for me when I want to introduce new players to the joy of video games than did the complicated games of the 1980s and 1990s. Since I became a full-time academic, my own life circumstances have also been changing. I now have meetings, papers to write, trips to make, and it has become harder to find the long stretches of time required for playing the large, time-intensive video games that I still love. Casual games just fit in better with my life.

One would think that making games that *fit* into people's lives was therefore the single most important problem that the video game industry had been working to solve. But in fact, the industry has spent decades solving an entirely different problem, that of how to create the best *graphics* possible.

The Problem with Graphics

[Microsoft on the Microsoft Xbox 360:] Microsoft Corporate Vice President and Chief XNA (TM) Architect J Allard further outlined the company's vision for the future of entertainment, citing the emergence of an "HD Era" in video games that is fueled by consumer demand for experiences that are always connected, always personalized and always in high-definition.[22]

[Sony on the Sony PlayStation 3:] In games, not only will movement of characters and objects be far more refined and realistic, but landscapes and virtual worlds can also be rendered in real-time, thereby elevating the freedom of graphics expression to levels not experienced in the past. Gamers will literally be able to dive into the realistic world seen in large-screen movies and experience the excitement in real-time.[23]

Upon entering the lecture hall for the Microsoft keynote at the Game Developers Conference in March 2005, I was handed a blue badge. Other attendees received yellow or black badges, but we did not know what their purpose was. The yearly Game Developers Conference is the place where the platform owners—currently Sony, Microsoft, and Nintendo—court developers and try to convince them to develop for *their* console. This was especially pertinent in 2005 since the then-current consoles (Play-Station 2, Xbox, and GameCube) were approaching the end of their lifetimes and developers were waiting for what would happen next. J Allard of Microsoft gave a conference keynote and proclaimed that the upcoming Xbox 360 would herald the coming of the HD era. The name *HD era* derived from the fact that the Xbox 360 would have graphics in *high definition*; it would show more pixels than earlier consoles. The Xbox 360 would also have other features such as the user's ability to connect to friends via the Internet, but *HD* was chosen as the moniker encompassing all of the experiences the console could give. At the end of the presentation, the audience was treated to a short animation showing a blue car, a yellow car, and a black car racing each other. The yellow car won, and the thousand attendees with correspondingly colored badges each won a high-definition television. This was Microsoft's take on what should define the next generation of video game consoles: higher definition graphics, more pixels. Sony was happy to follow suit, declaring that while HD really *was* the future, only the PlayStation 3 would be *true* high definition.[24] But not everybody at the conference was buying it. Game designer Greg Costikyan described his reaction like this: "Who was at the

Figure 1.7
Microsoft Xbox 360

Microsoft keynote? I don't know about you but it made my flesh crawl. The HD era? Bigger, louder? Big bucks to be made! Well not by you and me of course. Those budgets and teams ensure the death of innovation."[25] This was a good expression of the undercurrent of worry at the 2005 Game Developers Conference: the worry that developers would have to spend more resources creating game graphics, thereby pushing budgets to new heights at the expense of game design innovation.

In the then-upcoming generation of consoles (figures 1.7, 1.8, 1.9), the Nintendo Wii was the only one *not* promoted specifically on better graphics; in fact it did not even *have* the high-definition graphics that Sony and Microsoft were trumpeting. Figure 1.10 illustrates how the Wii is by far the technically weakest console of the generation,[26] but is also, as of February 2009, by far the most popular game console of the generation.[27] Technical selling points clearly do not drive sales of game consoles today.[28]

Figure 1.8
Sony PlayStation 3

Figure 1.9
Nintendo Wii (image courtesy of Nintendo America)

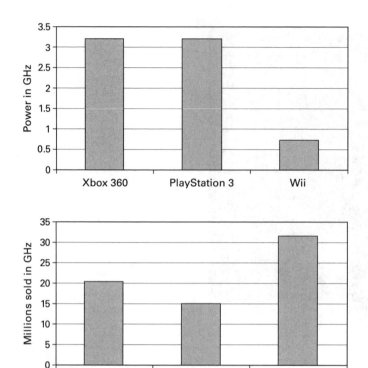

Figure 1.10
Power of game consoles compared to sales by February 2009

If the Wii lags in the graphical department, it does have a new kind of controller and a strategy for reaching a new, market of more casually oriented players. Judging from these numbers, the traditional way of selling new consoles and games via increased graphic fidelity has ceased to work[29]—or at least is beginning to be outshone by new ways of making games, and by more casual experiences aimed at more casual players.

From 3-D Space to Screen Space to Player Space

The problem with the industry focus on graphics technology is not that graphics are unimportant, but that *three-dimensional* graphics are not necessarily what players want. Casual game design is about making games fit in better with players' available time, but it is also about using space in a different way than one experiences in recent three-dimensional video

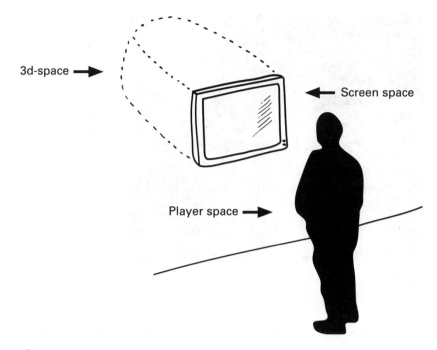

3d-space →

← Screen space

Player space →

Figure 1.11
3-D space, screen space, player space

games. Figure 1.11 shows how video games can involve three different types of space: whether sitting or standing, the player is situated in the *player space*, the physical space in front of the screen. The screen itself is a flat surface, the *screen space*. Any three-dimensional game presents a world inside the screen, a *3-D space*. (The real world of player space is of course also three-dimensional, but by 3-D space I mean the world projected by the screen.)

Early video games such as *Pac-Man* or *Pong*[30] were two-dimensional, but when games like *Wipeout*[31] (figure 1.12) were published in the early to mid 1990s, the then-amazing graphics looked like the future of video games, heralding that all video games would eventually become three-dimensional. Nevertheless, with casual games the history of video games took a different turn. The 1998 *Dance Dance Revolution* (figure 1.13) shifted the focus from 3-D space to the physical movement of the players on the game's dance pads. The game does feature a display, but most of the game's spectacle is in player space, the real-world area in which

Figure 1.12
Wipeout (Psygnosis 1995)

players move about. Furthermore, the 2004 downloadable casual game *Bejeweled 2 Deluxe*[32] (figure 1.14) is two-dimensional just like early arcade games. The movement to screen space and the movement to player space are core aspects of the trends in casual games that I will discuss in this book:

• *Downloadable casual games* are generally two-dimensional games that take place in screen space.
• *Mimetic interface games* are often three-dimensional, but encourage interaction between players in player space, and in such a way that player space and 3-D space appear continuous: when bowling in *Wii Sports*,[33] the game gives the impression that player space continues into the 3-D space of the game.

In short, video games started out as two-dimensional games on screen space, became windows to three-dimensional spaces, and now with casual games we see many games returning to both the two-dimensional screen space and to the concrete, real-world player space of the players. Casual games have a wide appeal because they move away from 3-D spaces, blending more easily with not only the time, but also the space in which we play a game.

Figure 1.13
Dance Dance Revolution player (Mario Tama/Getty Images)

Figure 1.14
Bejeweled 2 Deluxe (PopCap 2004)

Mimetic games move the action to player space, but many of them also encourage short game sessions played in social contexts. Such games, like all multiplayer games, are *socially embeddable*: games for which much of the interesting experience is not explicitly *in* the game, but is something that players add to the game. For example, if playing a competitive match of *Guitar Hero* or *Wii Tennis*, the game takes on meaning from the existing relations between the players. Playing a game against a friend, a significant other, a boss, or a child, adds meaning and special stakes to the game. Furthermore, people playing mimetic interface games are often themselves a spectacle, making these games more interesting even for those who are not playing.

Casual games are new, but new by reaching back in game history and by borrowing liberally from non-video-game activities. Video games are becoming normal; during the history of *all* games, everybody, young and old, has played games of one kind or another. The rise of casual games is the end of that small historical anomaly of the 1980s and 1990s when video games were played by only a small part of the population.

About This Book

This book is meant to capture *what is happening* with video games. In order to do that, I look at the games themselves, at players, and at developers. I will examine the designs of popular casual games, showing the common qualities that make them different from traditional hardcore video games. In order to learn about the habits and opinions of players, I have conducted a survey of two hundred casual game players. I have also made in-depth interviews with a number of game players and game developers.

This book is also meant to fill a void in the rapidly expanding field of video game studies. Most books on video games have tended to be either entirely general (such as Salen and Zimmerman's *Rules of Play*[34]), or focused on specific games (such as T. L. Taylor's *Play between Worlds*[35] on *EverQuest*[36]), or covering specific aspects of all games (such as Mia Consalvo's *Cheating*[37]). Here I am exploring a middle level of video game studies by looking at the position of casual games in the history of video games and games as such. My feeling is that video game studies must keep improving its tools—tools that must be more than general claims about all games and players, and more than the mere descriptions of single games or players. It is paramount that we can acknowledge player culture without treating games as black boxes, and we must be able to discuss game design without ignoring the players. We must be able to talk about how a single session of a small game is part of the entire history of games. This book constitutes my proposal for how this can be done.

Following this introduction, the rest of the book examines the casual revolution around two questions:

1. How did casual games appear, and how do they relate to the history of video games and nondigital games?
2. How do players and games interact? How do players engage with a given game?

Chapter 2 begins by combining these questions: the terms *casual games* and *casual players* are recent inventions, but they are a response to a time period during which video games became ever more complex and demanded ever more video game knowledge from a player. Casual game design, then, reinvents video games and goes hand in hand with a reinvention of the video game player. The casual revolution contains a new way for players and games to engage. Casual games share a set of design characteristics that I judge against common conceptions of casual and hardcore players, and show that while actual players are much more varied than can be expressed with the "hardcore" or "casual" categories, casual game design is successful because it is flexible toward different tastes and different usages.

As is often the case, painting a big historical picture makes it easier to perceive the details of what is happening now: chapters 3 and 4 consider casual games in a historical perspective. Chapter 3 shows Solitaire (or Patience) as a proto-casual game that became one of the most popular games played on computers because it was already familiar to players. Solitaire illustrates how a game is always perceived against the background of the games that a player has previously tried, and that the main barrier to playing video games has not been computer technology, but game design.

Chapter 4 focuses on history in a shorter time span: I examine the success of downloadable casual games and review the history of matching tile games. These often simple games evolve only gradually over time, which puts game developers in the treacherous position of having to differentiate themselves from previous games, while still building sufficiently on well-known game conventions that a game is easily accessible to new players. Developers of downloadable casual games borrow generously from earlier games, but they openly try to position themselves as innovative.

Chapter 5, 6, and 7 each tackle the ways in which players and games interact. Chapter 5 examines mimetic interface games, especially *Guitar Hero*, *Rock Band*, and games played on Nintendo Wii, to show that their success is due in part to the fact that they do *not* require players to know video game history, but build on more commonly known activities such as tennis and guitar playing. They are also often social games that move the game action into the space in which players play.

The interstitial chapter 6 explains why games can be social in the first place, by showing how even strategically shallow games like Parcheesi are considered social games, and how most of the meaning of such games is brought to the game by the players. Nevertheless, the meaning of a game is facilitated by design: when players can choose among playing to win, playing to keep the game interesting, or playing to manage the social situation, a game quickly become socially meaningful.

Chapter 7 asks why some games, such as *Guitar Hero, Sims,* or the *Grand Theft Auto* series are open to many levels of engagement and to being played in many different ways. These games are widely popular because they do not force the player to follow the goal. With this observation, the book returns to the question of history, showing that economical considerations meant that early arcade games *had* to punish players harshly for not reaching the game goal, thereby narrowing the range of available playing styles. Newer large-scale games are meaningful with both small and large time investments because the player is free to not follow the game goals.

Chapter 8 concludes the book by considering the skepticism that many traditional hardcore game players have toward casual games, asking whether game developers have an obligation to make games for people other than themselves, and placing casual games in the history of video games.

Finally, three appendixes document the habits and attitudes of casual game players and developers.

Appendix A contains the results of a survey of players of downloadable casual games.

Appendix B is a collection of player life stories gathered through the survey in appendix A and through additional interviews.

Appendix C contains excerpts from interviews with game developers about their views on the changes in video game design and in video game audiences.

2 What Is Casual?

A publisher of downloadable casual games gives the following somewhat sarcastic piece of advice to potential game developers: "Your game needs to be a casual game, appropriate for all ages. Remember, over half of this audience is women, and over 80% are over 30. As realistic as the blood effects are in your Vampire Corpse Feast game, it isn't going to sell to a casual audience and wouldn't be appropriate for Oberon Games."[1]

This paints a picture of an immature game developer that focuses on vampires and realistic graphics. Vampires and graphical realism—this, then is what casual games *are not*. Casual games are positioned as a rejection of traditional hardcore game design, with its gory themes and focus on technological capabilities. But what is *casual*?

The idea of casual games that reach casual players gained traction during the late 1990s, but it has a longer history that I will return to. The more recent history of the term *casual games* appears to begin at the 1998 Computer Game Developers Conference,[2] where puzzle designer Scott Kim described the dispositions of casual players: "The point is that people play different types of games for different reasons. Expert gamers [synonym for players] play for the longer term rewards of competition and rankings, whereas casual gamers play for the shorter-term rewards of beauty and distraction."[3]

While Kim denies having coined the term *casual* in relation to video games,[4] the talk also describes a trend called "games for the rest of us":

Most computer games are written for computer game hobbyists. Games like Tomb Raider, Quake and Dark Forces ... are epic combat games aimed at young males willing to invest dozens of hours developing complex battle skills. Each year the games get bigger and more technologically sophisticated. I call these sorts of games "games for gamers."

There is another rapidly growing segment of the computer gaming world that marches to the beat of a different drummer. Games like Myst, Monopoly, and Lego Island—also three of the most popular PC games of 1997—appeal to a much broader audience of males and females of all ages that want easy-to-learn family games. These games tend to use simple technology, and sell steadily year after year. I call this broad class of games "games for the rest of us."[5]

In the popular press, a 1999 *New York Times* article discussed the game *Deer Hunter*[6] as an example of a game for this new market: "The hunting games are also an early wave of what industry analysts call casual games, easier to play and more mundane in appearance than the special-effects-on-a-screen adventures preferred by what the industry considers hardcore gamers—male techies in the 16-to-35 age range. Computer Gaming World, the bible of computer game purists, once criticized Deer Hunter as 'boring' and 'repetitive' with 'lame' graphics."[7]

The terms *casual* and *hardcore* are, evidently, often used for positioning two categories of games against each other. The quote beginning this chapter painted hardcore games as vampire games with overly technical graphics, and casual games are sometimes dismissed as "boring" or having "lame" graphics. In the developer interviews in appendix C, Frank Lantz asserts that casual implies a "dumbing down" of games, and Eric Zimmerman argues that it entails a light and not-so-meaningful relation to a game: "As a producer of culture, I like to think that my audience can have a deep and dedicated and meaningful relationship with the works that I produce. And the notion of a casual game implies a light and less meaningful relationship to the work."[8] Neither Lantz nor Zimmerman are in any way opposed to making games for a broad audience, but they believe that "casual" connotes bland or shallow games.

While the term *casual* is sometimes controversial, it plays an important role in the changing landscape of video games. Let me therefore note that the idea of casual games has appeared specifically as a contrast to the idea that video games could only be made for a hardcore game audience.[9] The question then becomes: from where does the idea of a narrow, "hardcore" audience for video games come? Surprisingly, a 1974 press article introduced video games completely differently, emphasizing their "very nearly universal" appeal.[10] Another article states: "Never before has an amusement game been so widely accepted by all ages. Everyone from teenagers to senior citizens enjoy the challenge that the Video Games offer."[11]

While these early news stories present video games as appealing to "everyone," only a few years later video games were discussed with the assumption that they were intended for boys. A 1981 newspaper story singles out *Pac-Man* as a game that "surprisingly" appeals to women: "Midway, which licensed Pac-Man from a Japanese concern, is as surprised as everyone else by the game's appeal to women. 'We only became aware of it when women kept calling us and saying it was "adorable,"' Larry Berke, Midway's director of sales, said."[12]

In order to capitalize on this trend, Midway introduced *Ms. Pac-Man*,[13] functionally quite similar to *Pac-Man*, but with a feminized protagonist: "We've noticed a recent trend in our game pavilions that indicates a tremendous female acceptance of the Pac-Man game," says Castle Park marketing chief Michael Leone. "I guess it was only natural for Midway, manufacturer of the game, to introduce a Ms. Pac-Man. To woo the potential female video addict, Ms. Pac-Man is outfitted with more fashion wrinkles than a new Halston. Pac-Man is a homely little yellow critter on a screen, but his female video counterpart is resplendent in red lips and eyelashes, with a bow above her brow."[14]

During the 1970s video games apparently became increasingly associated with young men, but since then a number of video games have been declared *the* game that finally attracted new players. In the history of video games, a few particular games stand out for reaching a broad audience:

• The preceding 1981 quotations suggest that the first game to challenge the then-new young-male-gamer stereotype was the 1980 *Pac-Man*.
• It could also be argued that the 1985 *Tetris*[15] was the first casual game.[16]
• Steve Meretzky has argued that the first casual game was the 1990 Solitaire on Windows 3.0.[17]
• A study of U.S. baby boomer game players (born 1946–1964) showed the 1993 *Myst*[18] adventure game to be the most common *first* computer game played, as well as the most common favorite computer game.[19]
• The 2001 *Bejeweled* is sometimes hailed as "the genesis of the casual gaming boom."[20]

What is new today is not that a single game reaches a broad audience, but that a large number of games do, and that dedicated distribution channels and video game consoles are reaching beyond the traditional video game audience.

The Stereotypes of Casual and Hardcore Players

Prior to the launch of the Wii console, Reginald Fils-Aime from Nintendo described the company strategy as a mission to reach a broader audience by downplaying the importance of graphics and making games that were not "intimidating." At the same time, Fils-Aime said, Nintendo wanted to assuage fears that [hard] "core" game players would be left behind:

Q: What made Nintendo try to do something dramatically different with the Wii?
F-A: Our focus is interactive game play, a whole new way to play, that puts fun back into this business. It allows everybody to pick up and play and isn't focused on the core[21] gamer.
Q: The Wii seems to emphasize the controller, not heavy attention on graphics. Is that by design?
F-A: That is exactly by design. Our visuals for Wii will look fantastic, but in the end, prettier pictures will not bring new gamers and casual gamers into this industry. It has to be about the ability to pick up a controller, not be intimidated, and have fun immediately. The trick is being able to do that, not only with the new casual gamer, but do it in a way that the core gamer gets excited as well.[22]

Here, Fils-Aime uses a common description of casual players: as players who are not so interested in "graphics" as such, who have little knowledge of video game conventions and are therefore easily intimidated, and who desire "quick fun." Furthermore, games must be easy to learn not only because casual players prefer simple games but also because it is assumed they spend little time playing games: "These are games created for people that weren't sitting down for hours to play games," said Mr. Tinney of Large Animal. "They're taking a break from something they're doing and play for a few minutes." In contrast to the audience for hardcore games, he said, "women play these games as much or more than men do."[23]

Casual players are also generally assumed to dislike difficult games,[24] and it has even been claimed that a game for a casual audience can *never* be too easy.[25] This description of casual players has a mirror image of "hardcore gamers," who are often described as being committed to a game at any cost: "Do not let anybody disturb you. As soon as you come from school or work, immediately turn on your computer (or, better, do not turn it off at all) and load your favorite game. Do not answer phone or doorbell. Do not go to the bathroom at all—you could have done that at work! Newbies should play at least until midnight, advanced gamers need not sleep at all. On weekends, [hardcore players] must stay at their monitors non-stop."[26]

Even if this is an ironic quote, it identifies a *hardcore ethic*: spend as much time as possible, play as difficult games as possible, play games at the expense of everything else. Until recently, the game website Kotaku explicitly encouraged its readers not to "get a life": "As if you don't waste enough of your time in a gamer's haze, here's Kotaku: a gamer's guide that goes beyond the press release. Gossip, cheats, criticism, design, nostalgia, prediction. Don't get a life just yet."[27]

Let me sum up these stereotypical descriptions of casual and hardcore players. They should be understood as exactly that—stereotypes. As will become clear, most actual game players do not match these categories precisely:

• The *stereotypical casual player* has a preference for positive and pleasant fictions, has played few video games, is willing to commit small amounts of time and resources toward playing video games, and dislikes difficult games.
• The *stereotypical hardcore player* has a preference for emotionally negative fictions like science fiction, vampires, fantasy and war, has played a large number of video games, will invest large amounts of time and resources toward playing video games, and enjoys difficult games.

Figures 2.1 and 2.2 illustrate the difference between stereotypical casual and hardcore players on the four scales of fiction preference, game knowledge, time investment, and attitude toward difficulty.

To what extent do actual players match these stereotypes? The stereotype of the casual player tends to be thoroughly disproved when users are studied: according to a 2006 survey of players of downloadable casual games, two thirds of them played nine two-hour game sessions a week.[28]

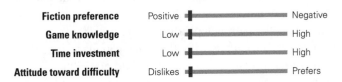

Figure 2.1
The stereotypical casual player

Fiction preference	Positive	Negative
Game knowledge	Low	High
Time investment	Low	High
Attitude toward difficulty	Dislikes	Prefers

Figure 2.2
The stereotypical hardcore player

My own survey of casual game players in appendix A also revealed surprisingly time-intensive playing habits. One player reports: "When I'm at home I usually play them for two or more hours a day. I will often put in an hour and a half in the morning, and then a shorter stint in the afternoon."[29] In terms of time investment, these players of casual games are much closer to the stereotype of the hardcore player. This raises two questions:

1. Do casual players really exist, or is the casual-player stereotype without any basis in reality?
2. Can all games simply be played in either hardcore or casual ways—however a player desires?

To answer both, I will first look at the *design* of casual games to see the kinds of engagement these games afford players.

The Elements of Casual Game Design

Comparing the design of the games I have described as casual—including downloadable casual games, *Guitar Hero*, and many Wii games—to more traditional video games yields five common casual design principles. To understand how these five work, think about how you use a video game over time:

1. First you see or hear about a game's *fiction* on the web, via the game's packaging, or from another source;
2. Then you learn to play the game, depending on its *usability*;
3. Next you try to match the game with the time you have available, depending on its required time investment and its *interruptibility*;
4. Then you continue to play the game if it has the right level of *difficulty*;
5. And finally, you continue playing if you like the content, the graphics, and the general *juiciness* (positive feedback) of the game.

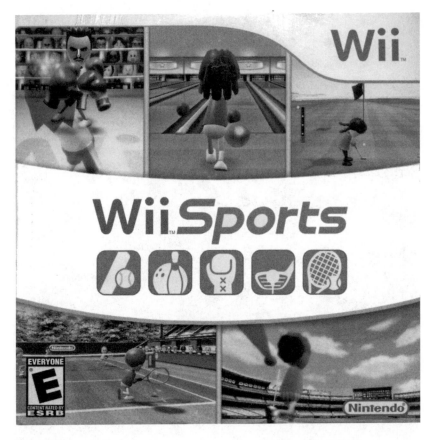

Figure 2.3
Wii Sports box art (Nintendo 2006)

Fiction

The first impression of a game comes from the presentation of what it is ostensibly *about*. There is a genuine difference between the setting portrayed on the cover of the casual *Wii Sports* (figure 2.3) and the cover of the hardcore *Gears of War*[30] (figure 2.4). The traditional hardcore game is often set in dangerous situations, allowing the player to take on the role of a soldier, or to crash cars, and so on. Casual games are often set in more positive and familiar settings. One could be tempted to say the sun always shines in casual games. In psychological terms, the fictions of the two games shown have different emotional *valence*, which refers to whether an emotion inclines you to approach something or to avoid

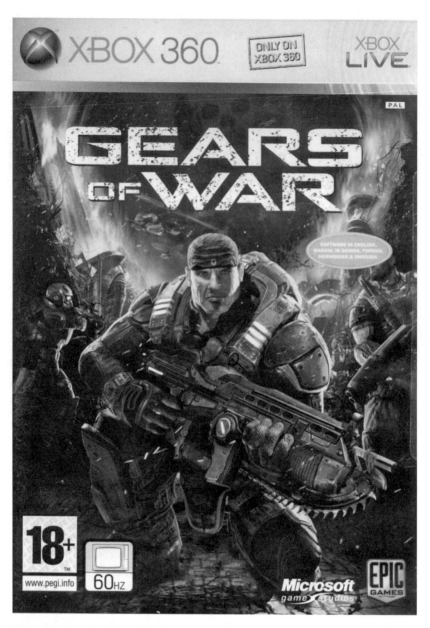

Figure 2.4
Gears of War box art (Epic Games 2006)

it. Chances are, if you walked down the street and encountered a typical casual game setting such as a restaurant or a tennis match, you would experience positive emotions and find the situation pleasant or attractive. On the other hand, if you walked down the street and encountered a typical hardcore game setting such as an armed conflict, you would in all likelihood experience negative emotions and perceive the situation as fundamentally unpleasant.[31] Casual games almost exclusively contain fictions with positive valence. Traditional hardcore games generally have fictions with negative valence, even if a few traditional hardcore games, such as hardcore tennis games, have a positive valence.

The cover of a game does more than signify a setting: if you are familiar with video game conventions, you can also use the cover to identify the genre of the game. The cover of *Gears of War* signifies a war setting to a non-player of video games, but to a player versed in video game conventions, the cover also signals that the game belongs to the genre of first-person shooters.

Usability

Downloadable casual games and mimetic interface games provide great improvements in user friendliness and usability compared to many traditional video games, but they do so in slightly different ways. Mimetic interface games work by creating new interfaces that build on conventions and activities from outside video games, while downloadable casual games use more traditional strategies from the field of usability and human-computer interface design[32] in order to make games easy to use.[33]

When I say that casual games are *easy to use*, it seems a paradox because games are also expected to be *difficult*. Why would usability even be an issue if we really want games to be challenging and difficult? The trick is that challenge and ease of use are parallel concerns: for example, a computer chess game has a *badly* designed interface if it is difficult to move the pieces, but the fact that it is difficult to win against the computer is *good* design. According to Michel Beaudouin-Lafon, interfaces are *interaction instruments* that mediate between the user and the *domain objects* the user wants to act on.[34] This explains the example of the chess game: while it is good design if the domain objects—the chess pieces— are strategically challenging, challenging instruments for moving said pieces would be bad design in a chess game. This is not the case for *all* games—think only of pick-up sticks or block removal games like *Jenga*[35]

Figure 2.5
Wii Sports tennis (Nintendo 2006)

or *Boom Blox*.[36] A given game can place a part, or all, of the challenge in the interaction instruments,[37] so the value of usability methods in video game design is in making sure it is the right parts of a game that are easy and the right parts that are challenging.

Following Beaudouin-Lafon, an interface can be evaluated on its *degree of compatibility*: this is a measure of the similarity between the physical action of the user and the action performed on the domain objects. For example, traditional tennis video games implement the action of serving in the pressing of a button at the right time when an energy meter on the screen peaks,[38] meaning that there is a low degree of compatibility between the player's actions and the events on the screen.

In Beaudouin-Lafon's terms, mimetic interface games have a high degree of compatibility, with the concrete actions of the player similar to the in-game actions: to hit the ball in *Wii Sports* tennis (figure 2.5), a player must swing his or her arm. This is similar (though not identical to) normal physical tennis. A high degree of compatibility is not the only thing that makes mimetic interface games work, as most of these games are about activities commonly represented in the media. When playing *Guitar Hero*, for example, players may not have any concrete experience playing guitar, but will have seen enough media representations of guitarists to know that one should place the left hand on the fret board and strum with the right hand. Mimetic interface games generally involve commonly known activities, and they have interfaces with a high degree of compatibility with those activities.

Downloadable casual games are played on personal computers designed for other purposes than playing these games, so usability must be achieved in some way other than by creating new game controllers. Going back to interface design literature, Ben Shneiderman recommends that interfaces should have

- continuous representation of the object of interest;
- physical actions (movement and selection by mouse, joystick, touch screen, etc.) or labeled button presses instead of complex syntax;
- rapid, incremental, reversible operations whose impact on the object of interest is immediately visible;
- layered or spiral approach to learning that permits usage with minimal knowledge. Novices can learn a modest and useful set of commands, which they can exercise till they become an "expert" at level 1 of the system. After obtaining reinforcing feedback from successful operation, users can gracefully expand their knowledge of features and gain fluency.[39]

Without mentioning such advice on interface design, the developers of the real estate game *Build-a-Lot*[40] (figure 2.6) reached the similar conclusion that objects should be continuously represented by making the game world no bigger than the screen:

One item we addressed early on was free scrolling. While the game was still a list of ideas on our white board, we decided that the gameplay area (i.e. the "map") was going to be fixed. Reducing the map to a single screen had many positive benefits for the project: 1) Development time was shortened because we did not need to develop a scrolling system. 2) A mini-map was not necessary which meant we would have more room for interface items. And, 3) the game would be much easier for casual gamers to learn since navigating a larger virtual world was not required.[41]

Build-a-Lot embodies many other principles for good interface design. To buy a house, the user can simply click on the house rather than on a separate menu. Furthermore, the game follows the layered approach mentioned previously, especially in the use of *combos*, allowing players to first play the game by way of simple actions, but subsequently awarding the player with bonuses for combining the basic actions in a special way. As for reversibility, games rarely allow players to undo an action, but

Figure 2.6
Build-a-Lot (HipSoft 2007)

downloadable casual game design structures difficulty and punishment so that the player can generally recover from making a mistake by performing well during the rest of a level.

Interruptibility

In a European study, non-players of video games reported that "lack of time" was their primary reason for not playing games.[42] Does that mean that casual game design reaches new players by introducing shorter games? The two aforementioned surveys of casual players revealed that they often play for long periods of time. Therefore the broader picture is that casual game design can reach new players by allowing them to play in short bursts, to interrupt a game and put it on hold, but without preventing players from engaging in longer sessions. This is the *interruptibility* found in casual game design, giving casual games flexibility in the time investment they ask from players. This flexibility has a purely functional component and several psychological components. Functionally, downloadable casual games have *automatic saves*, allowing the player to simply close the window of the game should the need arise, leaving the game within a few seconds. The next time the game is started, the player will be asked if he or she wants to continue the game. Compare this to more traditional console game design such as in *Grand Theft Auto: San Andreas*,[43] where the player must

1. reach a save point in order to save;
2. select a save slot;
3. confirm to overwrite the file;
4. wait for the save to happen.

Downloadable casual game design allows the player to enter and leave a game very quickly, making it possible to play a game while at work, for example, or while waiting for a phone call. One quarter of "white collar" workers play video games at work,[44] so there certainly is a demand for this type of flexibility.

Multiplayer mimetic interface games are not interruptible in the same way as downloadable casual games, but they are generally based on activities of short duration such as playing a song or a game of bowling. In the survey of casual game players discussed in appendix A, answering the phone was reported as the most common source of interruption (figure 2.7). When the game is played in a social situation, there is likely less need for interruptibility as there may be social pressure to *not* answer the phone.

Interruptibility also has the psychological aspect of whether the player is informed in advance of a game session's length. The player has such knowledge in *Guitar Hero* or *Rock Band* (figure 2.8) when choosing to play a single song. In the tennis game of *Wii Sports*, the player can choose the number of sets to play. Downloadable casual games often have maps showing the player how many levels it takes to complete the game. Compare this to traditional arcade games in which play time is determined by player skill, and to a game like *Gears of War*, in which the player has to guess the game's length from other cues. In other words, the perceived ahead-of-time time commitment of a game can be just as important as the actual time commitment.

The second psychological aspect of time is whether it *feels* appropriate to leave the game. This runs across all game genres and the casual/ hardcore distinction, but whenever a player has solved all pressing tasks in a game, it facilitates taking a break. Even if a game technically allows the player to take a break at any time, it is still important for the game to

Figure 2.7
Most frequent answers to the question "What typically interrupts you when you play?" (image created using http://wordle.net/)

Figure 2.8
Rock Band choice of song (Harmonix 2007)

provide break-facilitating moments when there are no outstanding tasks to solve.

Difficulty and Punishment

The stereotype of the casual player implies that casual game design should always be *easy*. This has, in fact, been described as a good rule for casual game design: "No casual game has *ever* failed for being too easy."[45]

Is this true? Do casual players like easy games? Catherine Herdlick and Eric Zimmerman have discussed how the original version of their game *Shopmania*[46] (figure 2.9) was criticized by players for being exactly that, too easy:

In the original version of Shopmania, we approached the first several levels of the game as a gradual tutorial that introduced the player to the basic game elements and the core gameplay. This approach was based on the generally held casual game wisdom that downloadable games should be very easy to play, and that the frustration of losing a level should be minimized. However, the problem with going too far in this direction is that the game ends up feeling like interactive muzak: you can play forever and not really lose, and the essential tension and challenge of a good game are lost.[47]

Like the issue of time, difficulty is an issue for which the common stereotype of casual players is at least partially false; it turns out that casual

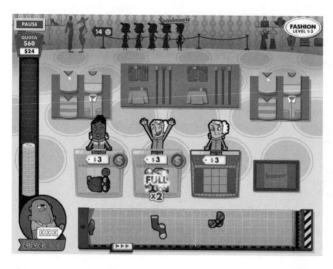

Figure 2.9
Shopmania (Gamelab 2006)

players often enjoy being challenged. When the game *Jewel Quest II*[48] (figure 2.10) came out, reviews described it as a *hard* game.[49] What did players think of this? One user described her relation to the game like this: "I love jewel quest 2. I [have] been playing jewel quest I for 4 years, and I think this game is very addictive and healthy, because it makes you think and is really fun!"[50]

It stands to reason that, having played the first version of the game for four years, this player had reached a point where she would enjoy a very difficult game. Casual game design must be usable, but the level of difficulty still needs to match the player's skills and preferences. Contrary to the stereotype, many players of casual games actively enjoy difficult games. A casual game player reported the following about her relation to difficulty in games: "I will quit any game that I can master in under ten minutes and doesn't introduce any more complications to the gameplay. Difficult games will frustrate me, but I'll keep playing."[51]

Another player of casual games described how she would retry a given level numerous times:

Q: Can you assign a number to how many times it is still enjoyable to retry a level before it becomes "too much," or does it depend on the game?

Figure 2.10
Jewel Quest II (iWin 2007)

A: I would say that if your skills have been honed by a steadily increasing learning curve, then ten to twenty would be the maximum replay number.[52]

For her, replaying a very difficult level in a game is perfectly acceptable and enjoyable, as long as the difficulty increase is reasonable. At the same time, she was willing to replay a level ten to twenty times—a high number by any standard.

In another common description, casual games are "easy to learn, but difficult to master."[53] That a game is difficult to master is sometimes referred to as *depth*, which means that players must continually expand their repertoire of skills[54] in order to progress in the game. A game that is too easy (like *Shopmania* discussed previously) does not require players to rethink their strategies or develop their skills. A separate study of players of a simple game showed that players who did very badly at that game gave it a low rating, and players who did somewhat well rated the game higher, but *players who never failed also gave the game a low rating.*[55] The ideal experience, for most players, seems to be failing some, and then winning. The experience of improving your skills, of gaining competence, is arguably at the core of almost all games, and those that do not provide that experience rarely become popular. Downloadable casual games, therefore, are not *easy* games; rather, they punish the player for mistakes in a slightly different way than the traditional hardcore game does. The real issue is not difficulty as such, but *how* the player is punished for failing.[56]

Consider a short history of difficulty and punishment in video games. Early arcade games were generally based on giving the player three tries to play the game, making it consistently more difficult as the player progressed, and forcing the player at the "game over" alert to start from the beginning (and insert new coins). When video games moved into the home in the late 1970s and early 1980s, they started to become longer affairs, but many retained a structure that forced a player to restart the entire game after running out of lives. For example, *Manic Miner*[57] (figure 2.11) has twenty levels that the player must complete with only three lives. Whenever game over is reached, the player must start over from the beginning. This is as different from casual game design as can be: the player is supremely punished for mistakes, the game is not in any way interruptible, and there is no way to save the game state and continue at a later point. The player is asked to commit hours of uninterrupted playing time in order to complete the game.

Another characteristic of a game like *Manic Miner* is that a level starts with exactly the same setup every time you enter it, so completing the game is a question of rote learning, of repeating a specific series of actions until you master it. This makes for a subjectively strong experience of being punished for failing in having to replay *exactly* the same early levels of a game over and over. Compare this to how failure works in *Magic Match: The Genie's Journey*[58] (figure 2.12): in the first picture, the game starts on level five. Subsequently, time runs out and level five restarts in the second picture *with a new random distribution of gems*. Downloadable casual games generally feature randomization, making the replaying of a level more interesting and less punishing. Another casual game, *Diner Dash*,[59] does not actually have randomization, but since the player rarely performs an identical set of actions at a given level, a level is experienced as nonidentical when replaying it.

The final feature that characterizes the difficulty and punishment structure of casual game design is that you rarely fail due to a single mistake

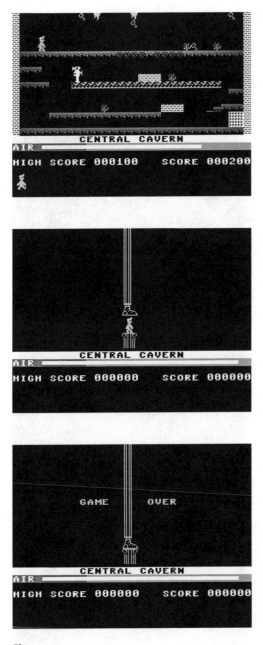

Figure 2.11
Manic Miner (Smith 1983)

Figure 2.12
Magic Match: The Genie's Journey (Friends Games 2007)

but rather to an accumulation of mistakes. Whereas you will instantly lose a life in *Manic Miner* if you touch an enemy, most downloadable casual games, as well as *Guitar Hero* only cause you to lose a life after you have accumulated a number of mistakes.

Excessive Positive Feedback: The Enigma of Juiciness

I watched a non-player of video games try the modern casual game *Peggle*.[60] When experiencing the end-of-level sequence shown in figure 2.13, he exclaimed, "It feels like somebody is praising me!" In discussing the usability of casual games, I described the design of a game interface as a question of *function*. But there is something else going on in casual games, something that is not predictable from the description of casual players. The end-of-level sequence of *Peggle* gives clear feedback to the player that he or she has completed the level, yet this is about more than *information*: the player already knows that the level is completed, so neither the "EXTREME FEVER" display, nor the extra bonus when the ball exits the bottom of the screen, nor the rainbow, nor the final extra bonus tally provide any new information to the player.

In the book *Emotional Design*,[61] design expert Donald Norman tells the story of how he changed his mind about the role of *beauty* in design: in his earlier work *The Design of Everyday Things*,[62] he had argued that the single most important attribute for an object like a teapot was its function, and how that function should always have a higher priority than secondary attributes like beauty. *Emotional Design* is about the realization that it *is* important how a teapot looks, and that Norman's earlier work on usability had underestimated the importance of this visceral level of design.[63] Concerning video games, independent game designer Kyle Gabler uses the term *juiciness* to describe the type of visceral interface that gives excessive amounts of positive feedback in response the player's actions: "A juicy game element will bounce and wiggle and squirt and make a little noise when you touch it. A juicy game feels alive and responds to everything you do—tons of cascading action and response for minimal user input. It makes the player feel powerful and in control of the world, and it coaches them through the rules of the game by constantly letting them know on a per-interaction basis how they are doing."[64]

Juiciness does not simply communicate information or make the game easier to use, but it also gives the player an immediate, pleasurable experience. Juiciness is tied specifically to feedback for the actions of players,

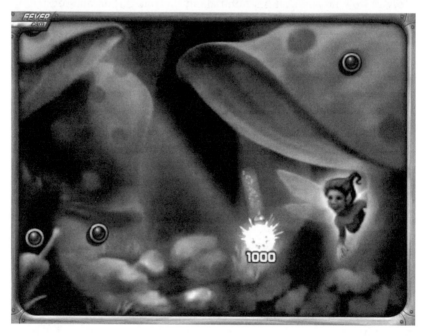

Figure 2.13
The juicy interface of *Peggle* (PopCap Games 2007)

Figure 2.13
(continued)

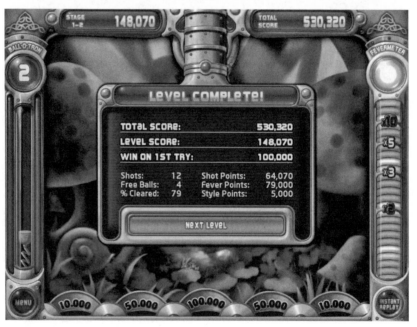

Figure 2.13
(continued)

Figure 2.14
Gears of War (Epic Games 2006)

seemingly enhancing the experience of feeling competent, or clever, when playing a game.

Juiciness is characteristic of casual game design, but does this mean that hardcore game design is not juicy? This is a bit more complicated. Consider *Gears of War*[65] shown in figure 2.14. *Gears of War* does have a large amount of juicy game elements: guns flare, things blow up. These game elements react excessively to player actions, demonstrating that juiciness is not exclusive to casual game design.

What is the difference? Juiciness in hardcore game design tends to take place in the 3-D space of the game as in *Gears of War*, but juiciness in casual games generally addresses the player directly. In the *Peggle* example, the first bonus sign, the "EXTREME FEVER" sign, the rainbow, and the final score tally are all elements in screen space. In film terms, hardcore game design has diegetic juiciness, which is juiciness within the game world, but casual game design is characterized by nondiegetic juiciness, which is juiciness that takes place outside the game world. Hardcore juiciness takes place in the 3-D space of the game; casual juiciness takes place in screen space, but addresses the player in player space.

Games and Players

To sum up, casual game design has five components.[66]

1. *Fiction*   The player is introduced to the game by way of a screenshot, a logo on a web page, or the physical game box. Casual games are generally set in pleasant environments. Casual game design has emotionally positive fictions as opposed to the mostly emotionally negative, "vampires and war" settings of traditional video games.

2. *Usability*   The player tries to play the game, and may or may not have trouble understanding how to play. Casual games presuppose little knowledge of video game conventions. Casual game design is very usable.

3. *Interruptibility*   A game demands a certain time commitment from the player. It is not that casual games can only be played for short periods of time, but that casual game design *allows* the player to play a game in brief bursts. Casual game design is very interruptible.

4. *Difficulty and punishment*   A game challenges and punishes the player for failing. Casual games often become very difficult during the playing of a game, but they do not force the player to replay large parts of the game. Single-player casual game design has lenient punishments for failing. The experience of punishment in multiplayer casual game design depends on who plays.

5. *Juiciness*   Though this was not predicted by the description of casual players, casual game design commonly features excessive positive feedback for every successful action the player performs. Casual game design is very juicy.

If the stereotype of the casual player is someone who only likes mainstream fictions, who has little knowledge of video game conventions, who is willing to invest little time in playing games, and who is averse to difficulty in games, then we must conclude that this stereotype has been thoroughly shattered on three of four counts. The downloadable casual game players of the study in appendix A exhibit much knowledge about the games they play and invest much time in playing. Many of the surveyed players report a preference for difficult games; more players consider it worse for a game to be too easy than for it to be too hard. These players are better illustrated in figure 2.15, as having acquired a large amount of knowledge about game conventions in casual games, being willing to invest large amounts of time in playing games, *and* having a surprisingly high tolerance for difficulty in games.

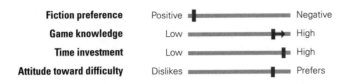

Fiction preference	Positive	▬▬▬▬▬▬▬▬	Negative
Game knowledge	Low	▬▬▬▬▬▶	High
Time investment	Low	▬▬▬▬▶	High
Attitude toward difficulty	Dislikes	▬▬▬▬	Prefers

Figure 2.15
Interviewed casual players

Studies of players of downloadable casual games tend to foster surprising headlines such as "'Casual' Players Exhibit Heavy Game Usage."[67] My own survey has similar results. As discussed in appendix A, such headlines should be taken with a grain of salt: the interviewees volunteer to answer the surveys, so it is predictable that avid players are more likely to respond than other players. There may be players who are closer to the stereotypical casual player discussed earlier, but they will have little motivation for answering questionnaires about casual games.[68] Casual game design *does* lower the barriers to entry significantly for new players, and it does provide a range of possible experiences. This means that even if the stereotype of the casual player does not describe actual players very well, it is still valuable for understanding how game designs can help players integrate a game into their lives. Table 2.1 shows how the elements of casual game design each support the characteristics of the stereotypical casual player.

Casual game design is also valuable because stereotypical hardcore players are under threat. Not, perhaps, as a cultural group, but as individuals. A common complaint is that a life with children, jobs, and general adult responsibilities is not conducive to playing video games for long periods of time. The player that at one time was a stereotypical hardcore player may find him or herself in a new life situation: still wanting to play video games, but only able to play short sessions at a time. Many players of casual games are such ex-hardcore players as illustrated in figure 2.16: they probably still have the same taste in fiction, but are unable to invest large amounts of time in playing games. Consequently their knowledge of video game conventions becomes dated. Presumably they also have a lower tolerance for difficulty so as to be able to make progress in or complete a game within the time they have for playing games.

We can individually switch between being casual players and hardcore players. This can happen over a long period of time as you gain more

Table 2.1
Casual players and casual game design

Stereotypical casual player	How casual design supports the stereotypical casual player
Preference for positive fictions	**Positive fictions** support players with such taste.
Little knowledge of game conventions	**Usable design** supports players with little knowledge of game conventions. **Juiciness** gives players constant feedback about how well they are doing.
Low time investment	**Interruptibility** allows *both* playing in short bursts with little time investment *and* playing with large time investments. *In multiplayer games, this depends more on other players than on the game design as such.*
Low difficulty tolerance	**Lenient punishments** for failure: casual game design generally *does not support players who dislike difficulty entirely*, but supports players who dislike replaying large parts of a game over, and who dislike rote learning. *In multiplayer games, this depends on the social consequences of losing, as discussed in chapter 6.*

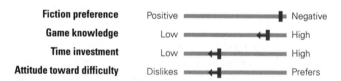

Figure 2.16
Ex-hardcore players

video game knowledge, or it can happen if you find yourself in changed life circumstances. We can also switch on daily basis: the fiction preferences that we express will surely change depending on who we play with. Our willingness to invest time depends on where we are playing and how busy we are. Our tolerance for difficulty depends on our mood. We each have certain dispositions toward games, but we also change.

At this point, it seems pertinent to ask the following question: *Which is it?* Am I saying that the casual revolution has everything to do with game design, or am I saying that it has everything to do with casual players? In chapter 1 I discussed why it is tempting to try to understand the casual revolution by focusing on either games or players, but I also argued that it would be wrong to make such a choice. While video games have only become a major subject of serious study at universities within the last ten years, much of that time has been a tug of war between those who advocate looking at games and those who advocate looking at players. This has led to the problem that any examination of game design can be criticized for ignoring players: one theorist denounces any focus on games as "unsustainable formalism."[69] Conversely, any researcher who focuses on examining players is vulnerable to being criticized for spending too little time examining and, indeed, playing games.[70] This is a competition between *player-centric* and *game-centric* views of how games and players should be understood. Let me briefly demonstrate both why this is a genuine discussion and why it is important to get beyond it.

• *Player-centric view*[71] If I start with casual players and focus on the way they play, it seems that players can take a video game and use it in any way they want. Yet different players enjoy different games, and casual game design supports more different ways of playing than traditional video game design does, hence casting doubt on a purely player-centric view.

• *Game-centric view*[72] If I start with casual games and focus on their design, it seems clear that the casual revolution discussed here is first and foremost a question of new, more casual designs. Yet it turns out that many players of casual games play in ways that appear distinctly non-casual, hence casting doubt on a purely game-centric view.

The better solution is to see how a game can be *more or less* flexible toward being played in different ways, and a player can be more or less flexible toward what a game asks of the player. Hardcore game design provides an inflexible ultimatum toward the player, asking him or her to commit much time and many resources to playing, but casual game design asks for *small* commitments while flexibly allowing the player to spend more time with the game if desired. Whereas the stereotypical hardcore player is strongly interested in games and *flexible* toward investing time and resources into playing games, the stereotypical casual player is only opportunistically interested in games and *inflexible* toward investing time and resources into playing games. This way it becomes clear that *hardcore* and *casual* have opposite meanings for games and players, as illustrated in table 2.2.

Table 2.2
Casual games and hardcore players are flexible, while hardcore games and casual players are inflexible

	Casual	Hardcore
Games	Flexible	Inflexible
Players	Inflexible	Flexible

Fiction preference
Game knowledge
Time investment
Attitude toward difficulty
Positive
Negative
Low
High
Low
High
Dislikes
Prefers

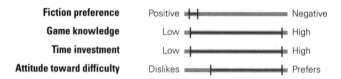

Figure 2.17
The affordances of casual game design: many types of game players are supported

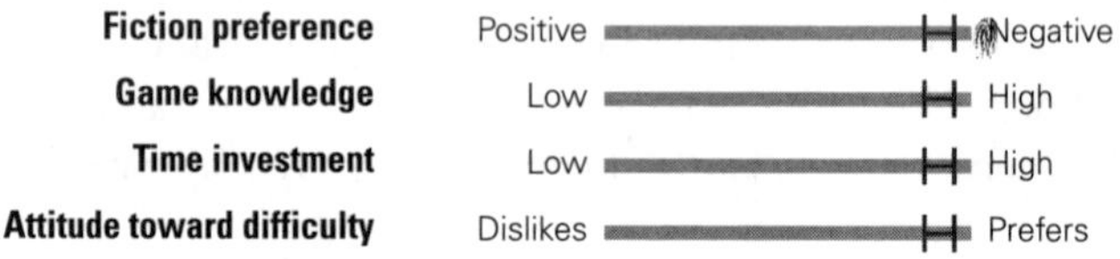

Fiction preference
Game knowledge
Time investment
Attitude toward difficulty
Positive
Negative
Low
High
Low
High
Dislikes
Prefers

Figure 2.18
The affordances of hardcore game design: only one type of game player is supported

Figure 2.17 shows how the flexibility of casual game design opens video games to a range of players: the high usability, high interruptibility, and lenient difficulty/punishment structures of casual game design means that casual games do not ask that much of players in terms of resources, game knowledge, or time investment. Casual *games* are consequently flexible toward being played by many players in many ways, but the stereotypical casual *player* is inflexible toward accommodating the demands of a game. Even if the fiction of casual game design tends to be unapologetically positive, casual game design still opens the games to players with little or a lot of game knowledge, and for playing with little or a large time investment. Contrary to popular belief, casual games are only rarely open to players with no tolerance for difficulty.

Figure 2.18 shows how hardcore game design is inflexible in that it asks for many resources from the player, and requires much knowledge of game conventions, much time investment, and a preference for diffi-culty. (Chapter 7 discusses the counterexample of hardcore games that al-low players to not follow the game goal and thereby gives them the option of playing in more casual ways.) Conversely, a stereotypical hardcore player is flexible as to accommodating what a game asks for. Hence, as long as the game has sufficient depth it is not a problem for a stereotypi-cal hardcore player to play a casual game since the game does not ask for something that the player will not deliver, but as a general rule it is a problem for a stereotypical casual player to play a hardcore game.

Is This a Casual Game?

With the elements of casual game design identified, it becomes easier to discuss specific games. It is rarely the case that a game is *either* casual or hardcore, but the design elements of a given game can pull it in either direction.

Chess
Let me start by discussing the traditional board game of chess. Chess is played in many different ways: it can be a somewhat relaxed game, or it can be a highly competitive game to which some players dedicate their entire life. In that case, does or doesn't chess embody the elements of ca-sual design that I have discussed in this chapter?

• The fiction of chess is a conflict between two societies, but it is probably not the reason why people play the game. Still, this fiction has likely been more resonant with players historically than it is now.
• Chess is fairly complicated compared to, say, checkers, but it is a mod-erately usable game compared to more complicated video games.
• The interruptibility and time investment components of chess depend almost exclusively on who it is played with, and in what context: a game of speed chess is a short, intensive experience but chess played by mail takes a lot longer and is less intensive. The player who merely wants to play a game with friends or family on occasion has to invest only a little time; the player who wants to compete on a tournament level must dedi-cate major portions of his or her life.
• Difficulty and punishment for failure depends on the player's stake in the social situation the game is played in. Replay is never identical.
• While chess exists in many forms and implementations, neither analog nor digital versions of the game tend to have much juiciness.

Chess, then, is a moderately usable game for which the time investment and the difficulty depend almost exclusively on what the player is trying to achieve and who the player is playing against. I cannot answer the question of whether chess is a casual game, as that is determined by the context in which chess is played.

Guitar Hero

Guitar Hero is similar to chess in that it serves two very different functions: it can be played as a relaxed social game, and it can be played as an intensive game by players who wish to master the game on expert difficulty setting or to partake in competitions.

• The fiction of *Guitar Hero* refers to a well-known stereotype of over-the-top rock music styles and musician poses.
• The basic activity of playing guitar is well known by players, and the mimetic interface mimics that activity with a high degree of compatibility, making the game very usable.
• The time investment can be either low or high depending on what the player is trying to achieve. The game can technically be paused at any time, but in terms of fitting into a player's life, the moderate length of a game session is more important.
• Concerning difficulty, *Guitar Hero* sets a relatively low bar for completing a given song while presenting a range of measurements of how well the player performs. This means that if a player only is interested in playing a song on a low difficulty setting, he or she is not forced to replay that song. On the other hand, if the player wants to complete the game, play on higher difficulty levels, or achieve a high score, the game punishes failure very harshly by forcing the player to replay a song until it has been mastered.
• *Guitar Hero* is an extremely juicy game, featuring large amounts of positive feedback for everything from the isolated action of pressing a single button, to correctly hitting a long sequence of notes, to crowds cheering the superior performance.

In other words, *Guitar Hero* has basic similarities with chess in that it matches casual game design principles if the player is only trying to play a song or two without worrying about achieving a high score or completing the game. If the player tries to complete the game or achieve a high score, the time investment and difficulty tolerance required from the game become decidedly non-casual. Chapter 7 discusses this flexible quality of *Guitar Hero* and other games in more detail.

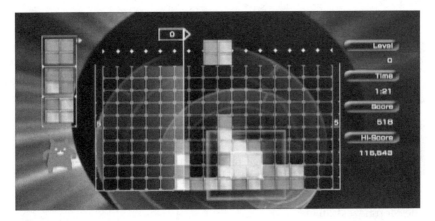

Figure 2.19
Lumines Live! (Q Entertainment 2006)

Lumines Live!

Lumines Live![73] (shown in figure 2.19) is the Xbox 360 version of the Sony PSP puzzle game *Lumines*.[74] The game has a certain *Tetris*-like simplicity in its basic structure: players manipulate the falling tiles so that tiles of the same color are aligned in a 2×2 grid. But is *Lumines Live!* a casual game? Consider these factors:

• *Lumines Live!* is an abstract game whose futuristic or technological style falls somewhat outside the positive/negative fiction scale.

• *Lumines Live!* is simple enough to be very usable, but there is a caveat: although neither *Lumines!* nor *Lumines Live!* is a downloadable casual game played on a personal computer, conventional wisdom in that distribution channel is to never make the player control a game with anything but the mouse.[75] Since *Lumines (Live!)* is played with the game controllers of either the Xbox 360 or the PSP, this is at odds with that conventional wisdom, possibly alienating many potential players. Furthermore, for reasons that escape us, *Tetris*-like games with falling blocks have rarely done well in the downloadable casual games channel.[76]

• In terms of time investment, the basic "challenge" game mode consists of only one level with no save function, often requiring a player to play for thirty minutes or more. Thereby the interruptibility factor becomes very low and the required time investment becomes very high.

• Failing in the game forces the player to start over from the very beginning, so *Lumines Live!* has a very harsh punishment for failure.

• Speaking for its status as a casual game, *Lumines Live!* is extremely juicy, with much positive feedback for the basic manipulation of the falling blocks as well as for both normal and special matches.

Though *Lumines Live!* on the surface appears to be a simple game, it has a surprising amount of depth, with the instructions including a range of tips for advanced playing strategies. This is the flip side of the question of time investment: a game can afford the player a meaningful experience with a short time investment, but does the game continue to be a meaningful experience with a *large* time investment? The depth of *Lumines Live!* affords exactly that, intensive playing with a large time investment by allowing players to continue to refine their skills at the game and to develop new strategies.

Is *Lumines Live!* a casual game? Though *Lumines*, the original Sony PSP handheld version of the game,[77] is almost identical, it is subtly different due to the constraints of the platform: where the Xbox 360 version forces the player into playing an entire game in one setting, the built-in pause function of the Sony PSP allows a player to pause the PSP and resume the game days or weeks later. The PSP version of the game is much more flexible, and interruptible, *but this is due to the design of the platform, not the design of the game.*

Lumines follows casual game design to the extent that it is a very simple and usable game, but it does not follow casual game design in that it is played with a standard game controller; requires a large, uninterrupted time investment when played on the Xbox 360, and has a harsh punishment structure. The move from the Sony PSP to the Xbox 360 made *Lumines* less casual.

Wii Sports

I have already singled out *Wii Sports* as an example of a casual game that has reached a broad audience. As of 2008, *Wii Sports* is the bestselling Wii game,[78] though it should be said that it came for free with Wii consoles purchased outside Japan. This is why I describe *Wii Sports* as a casual game:

• The fiction of *Wii Sports* is a cartoony and friendly version of the commonly known sports of tennis, bowling, golf, boxing, and baseball. In other words, the game occupies the very positive end of the fiction scale.
• Since players have seen these sports performed before, and since *Wii Sports* features a mimetic interface, the game is very usable.

• The required time investment of *Wii Sports* is quite short—down to the length of a few sets of tennis or a game of bowling.
• Since the individual games of *Wii Sports* are short games that you play many times against other players, the difficulty of the games and the consequences for failing are decided by dynamics of the group that is playing.
• *Wii Sports* is a fairly juicy game by way of feedback for individual actions such as hitting a ball. When completing player turns or entire games, the characters in the game provide a different kind of juiciness by expressing joy or dismay in responses to success or failure.

Wii Sports is similar to *Guitar Hero* in being a relaxed social game, but unlike *Guitar Hero* it does not work very well as a time-intensive game. Critical reviews of *Wii Sports* have focused particularly on the perceived lack of depth,[79] implying that intensive playing of the game is not rewarded by improved skills. Yet this is exactly what makes it a relaxed social game: popular social games like *Parcheesi* or *Monopoly* generally have large amounts of chance and moderate amounts of depth, making sure the same player does not win every time.

World of Warcraft

As of 2009, *World of Warcraft*[80] (figure 2.20) is the most popular massively multiplayer role-playing game in the Western world. *World of Warcraft* is famous for its time intensity, where reaching the initial maximum level sixty in the game takes an average of the equivalent of two full work months.[81] Here are the key characteristics of the game:

• The fiction of *World of Warcraft* is a somewhat generic fantasy world following role-playing game and fantasy conventions.
• While *World of Warcraft* is by no means a simple game, it is relatively usable compared to other games in the genre.[82]
• As noted, the required time investment to play *World of Warcraft* is immense, and the interruptibility of the game is minimal especially when playing in groups. *World of Warcraft* does mitigate this somewhat by allowing players to solo and play part of the game alone.
• Death in *World of Warcraft* is less punishing than in other and earlier games in the genre, but a player trying to complete a given quest in the game is still forced to replay that quest. However, the replay of a given quest is rarely identical.

Figure 2.20
World of Warcraft (Blizzard Entertainment 2004)

- Though nowhere on the level of *Guitar Hero* or *Lumines Live!*, *World of Warcraft* is moderately juicy, giving much positive feedback for fundamental actions such as casting spells and attacking.

World of Warcraft is a game that requires a decidedly non-casual time investment to play, but at the same time the game's popularity is partially due to it being slightly more usable and interruptible than other games in the genre. *World of Warcraft* is not a casual game, but it borrows some design principles from casual games.

Super Mario Galaxy

The 2007 *Super Mario Galaxy*[83] (figure 2.21) is an example of a Nintendo Wii game that cannot be described as a casual game for the following reasons:

- The fiction of *Super Mario Galaxy* has a colorful cartoony look, but also involves emotionally negative fiction and events such as monsters and the kidnapping of a friend (Princess Peach).

Figure 2.21
Super Mario Galaxy (Nintendo EAD Tokyo 2007), image courtesy of Nintendo America

- *Super Mario Galaxy* is primarily controlled via the control stick of the "nunchuck" controller, and the game takes places in 3-D space. This makes the game somewhat inaccessible to players who are not comfortable with this control method or with three-dimensional games as such.
- Completing the game takes ten to twenty hours, making it moderately time intensive.
- The difficulty and punishment structure of the game has three aspects: First, the game is structured around a number of smaller levels where the player who loses a life has to replay the *level* from its beginning, which is a harsh punishment structure. Second, the game is somewhat lenient in that the overall game gives the option of skipping a few levels if the player dislikes them or finds them hard. Third, though the player has a limited number of lives, losing all of his or her lives still only forces the player to replay the last level, rather than the entire game.
- While the difficulty and punishment structure of *Super Mario Galaxy* speaks against its status as a casual game, it has high levels of juiciness, with many small elements of positive feedback for basic actions as well as for completing levels in the game.

Regardless of being a game for the Nintendo Wii, *Super Mario Galaxy* is quite removed from the principles of casual game design I have outlined previously: its low usability, large time requirement, and generally harsh punishment structure make for a game that is not very casual, regardless of the platform it is made for.

Who Is Casual?

Wrote one hardcore reader to *PC Game Magazine*: "I read your 'Top 8 of 2008' feature in the January issue, and I was shocked and disappointed to find Peggle in there. Now, I know it is addictive, just like a ton of other low-rent Flash games. But to name it one of the best PC games of 2008 means you've either given up on the PC as a real gaming platform, or you've lost your sanity and can no longer be trusted to review videogames. So, which is it?"[84]

This hardcore player was unhappy to see his game magazine name *Peggle* as a top game of 2008. So far I have been telling the happy story about how video games reach new players. The other story is that of traditional hardcore players who fear their medium of choice will lose the qualities they enjoy. The two stories position casual games as either the salvation or the dumbing down of video games. Some elements of casual game design are certainly at odds with what we can call the hardcore ethic discussed earlier: that a game should be as challenging *as possible*, and that there is honor involved in spending as much time as possible with a game.

Conversely, some players describe themselves as casual players in order to distance themselves from the hardcore players assumed to spend excessive amounts of time on playing. For example:

When I call myself a "casual gamer," I mean someone who just plays for leisure, who doesn't devote a tremendous amount of time to playing. I knew people in college for whom gaming was a way of life: they would miss sleep to play, they would skip classes to play, and some of them would rather play games online than hang out with people in real life. Those are "hardcore" gamers, to me. . . . I consider my own habits casual because, among other things, my sense of identity isn't at all tied to my gaming ability (which is a good thing—I'm not very good at these games). I just play to amuse myself from time to time, and honestly if a game gets too hard I lose interest—I play to relax, not to be frustrated.[85]

As this chapter has demonstrated, casual game design gives a game a flexibility that allows players to use it in different ways. A well-designed casual game with sufficient depth can be enjoyed with both small and large time investments. For some players, casual game design can be a way *in* to video games, giving an opportunity to play video games in short sessions, while still allowing players to subsequently invest more time into playing. For formerly hardcore players with less time on their hands, casual game design grants them the opportunity to keep playing video games, even under changed life circumstances.

The casual revolution is a reinvention of both games and players: casual game design is a genuine innovation in game design and a return to lessons long forgotten, while the idea of the less-dedicated, less-obsessed casual player helps us to move beyond the prejudice that video game player are nerdy and socially inept. This lets developers reconsider who will be playing their games, when and why. It also removes some of the stigma that has been attached to video games, making it easier for us all to say that, "yes, I play video games."

What do you see in the image of *Puzzle Quest*[1] shown in figure 3.1? What do you think you *do* in this game? Do you have strategies for playing this game? Your answers depend on what games you have already played; on your knowledge of game conventions. Looking at the screenshot, you may or may not feel the pull discussed in chapter 1. You may or may not know what to do in the game, and you may or may not *want* to do it.

At the time of release, reviewers described *Puzzle Quest* as a combination of elements from matching tile games (games such as *Bejeweled* shown in figure 1.14), hitherto considered a casual game type, with role-playing game elements and a fantasy setting hitherto considered a stable of hardcore video games. The website *ign.com* specifically warned its hardcore readers not to be "ashamed of" *Puzzle Quest*'s use of casual game elements:

Hardcore gamers: be not ashamed of your love for jewel swapping. Casual gamers have their Bejeweled, and hardcore players have their RPGs[role-playing games]. Long have the two groups been content to remain separate and play their respective games. But the folks at D3 Publisher have begun a socialization experiment that may find gamers from both camps playing the same game. Puzzle Quest: Challenge of the Warlords attempts to marry characteristics of traditional RPGs with the pick-up-and-play mechanics of a casual puzzle game—and succeeds.[2]

Conversely, the casual game review site *Gamezebo* encouraged their readers not to be turned off by the fantasy and role-playing elements of *Puzzle Quest*, assuring that it was "a casual game at heart": "Classical fantasy trappings and detailed role-playing infrastructure aside, *Puzzle Quest: Challenge of the Warlords* proved a surprise hit on consoles like the Nintendo DS and PlayStation Portable for one simple reason—it's a casual game at heart. So don't be so quick to dismiss the outing, just because

Figure 3.1
Puzzle Quest (Infinite Interactive 2007)

you're not the sort who usually appreciates the complexity of *Dungeons &* *Dragons*-style romps or balks at the thought of playing swordsman or spellcaster."[3] Some user reviews show players making similar observations: "If you love RPGs and Puzzle games you will love this game!"[4]

How does this compare to the original intentions of the developers of the game? According to a conference presentation, *Puzzle Quest* was born of the idea of mixing two genres: "The inception of the game concept was a 'happy accident' caused by the process of iterative design. Steve started his plans with the ideas that he really liked *Bejeweled* and he really liked RPGs. Putting the two together seemed to result in a style of game that landed in that sweet spot the studio was aiming for, and seemed to be something compelling enough to play."[5]

This is the harmonious picture of innovation in games: a game developer has the intention of innovating; reviewers and players understand the intention and enjoy the game. Does it always work this well? If you look a little further, many users voice their frustrations with *Puzzle Quest*, as in these comments from an unhappy buyer:

I bought this game because I was looking for a game resembling *Bejeweled*. I think I would have rather found the real *Bejeweled* game. Wish Nintendo had made it . . . and *Atlantis* as well.

But, with the highest of hopes, I bought this game. Even with the manual, I'm still trying to figure out what the heck the purpose of this game is?

It sorta has some *Bejeweled* features . . . and a whisper of a story line, but other than that . . . I'm confused.

I enjoy the "practice" rounds where I can do a *Bejeweled* puzzle, but when it comes to combat with an opponent, I find myself sitting there while the opponent takes all the turns and completely stomps me. Then the game tells me I'm defeated . . . well . . . yeah!![6]

The player was apparently not familiar with the role-playing conventions in the game and was surprised by the spells and special objects *Puzzle Quest* added to the matching tile game formula. This prevented her from enjoying the game. Game conventions are double-edged swords: they are shorthands that allows games to build on other games, but they risk alienating users unfamiliar with those conventions.

Games that copy other games wholesale are derogatorily referred to as *clones*, but developers face a genuine challenge trying to strike a balance between innovation and cloning: on the one hand, players perceive new games on the basis of games they already know, and this puts pressure on developers to create games that are similar to previous games in order to give players an initial experience of competence. On the other hand, the player needs a reason to buy a *new* game, and there is therefore pressure on the developer to provide new, innovative experiences. On top of that, within the game development community, innovation has higher status than cloning, as will be documented in chapter 4.

This chapter examines how the proto-casual Solitaire card game became one of the most popular digital games in part because it was already well known by players. Solitaire is an example of how all games are created and used in four different time frames: historical time, design time, player lifetime, and game-playing time.

Genres, Fictions, and Interfaces

I described *Puzzle Quest* as a combination of two genres: role-playing games and matching tile puzzle games. (The status of matching tile games is discussed in further detail in chapter 4.) The challenge of talking about genre is that there are no clear agreements about how to define any given genre, and genre categories change over time.[7] Do players discuss and use genres for understanding games, or do only game developers and theorists? Figure 3.2 shows the most common words and phrases to appear in my survey of downloadable casual game players answering the question: What are your favorite casual games? While some of the words filled in refer to specific games (*Chocolatier*,[8] *Azada*[9]) and others are common parts of game titles (*dash*, *mania*), the most commonly used words and phrases refer to genres: *hidden object games*, *time management games*. The players surveyed here demonstrate a keen use of genre labels to describe the games they play, meaning that they see individual games as part of larger groups of games, and that they are aware of differences and similarities between games.

Game designer Greg Costikyan has pointed out that game genres are tied to game *mechanics*[10]—after what you *do* as a player, rather than after the fiction.[11] For example, to *match* tiles of similar color is a mechanic (as discussed in chapter 4); to *jump* in a platform game is a mechanic; to be able to *capture* the pieces of an opponent in *Parcheesi* is a mechanic. There are no game genres labeled "fantasy" or "science fiction," but there are strategy games, adventure games, puzzle games, role-playing genres, each of which is centered on what the player can do, the mechanics of that genre. Though Costikyan's claim that video game genres are defined by game mechanics is generally correct,[12] specific genres also have affinities to specific fictions: strategy games are generally tied to warlike fictions; matching tile games generally have bright and positive fictions; massively multiplayer games mostly have fantasy fictions. In addition, every genre has affinities with certain interface conventions. That is, although genres are named after game mechanics, they are also associated with other game elements, and all elements are potentially relevant to how a player understands a game. Figure 3.3 is an analysis of *Puzzle Quest* where the individual elements of the interface have been called out to identify the genres from which they derive. It is probably impossible to perform such an analysis exhaustively. For example, *Puzzle Quest* has turns, but where do turns come from? *Puzzle Quest* is also a game, and games as such contain a large number of conventions.[13]

A Short History of Solitaire Card Games

Consider the case of Solitaire or Patience card games such as *La Belle Lucie* shown in figure 3.4. Solitaire is an example of a popular game that

Figure 3.2
Most frequent answers to the question "What are your favorite casual games?" (image created using http://wordle.net/)

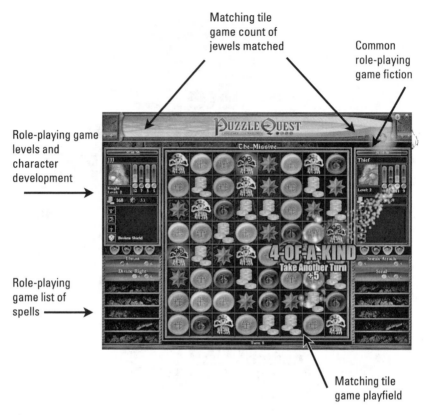

Matching tile
game count of
jewels matched

Common
role-playing
game fiction

Role-playing game
levels and
character
development

Role-playing
game list of
spells

Matching tile
game playfield

Figure 3.3
Identifiable game conventions and their sources in *Puzzle Quest* (Infinite Interactive 2007)

has developed over considerable time and has crossed technological boundaries. The first known references to Solitaire are from Germany and Scandinavia around the year 1800, with the first collection published in Moscow in 1826, and Solitaire gaining popularity in Europe during the later part of nineteenth century.[14] Solitaire has undergone a significant resurgence in popularity after becoming available on modern computers. The ease by which Solitaire became a computer game demonstrates the importance of player familiarity with game conventions. The history of Solitaire also reveals that, like video games today, Solitaire has been associated with a specific audience.

In English, the 1876 *Lady Cadogan's Illustrated Games of Patience*[15] was the first collection of Solitaire card games. Twenty-two years later,

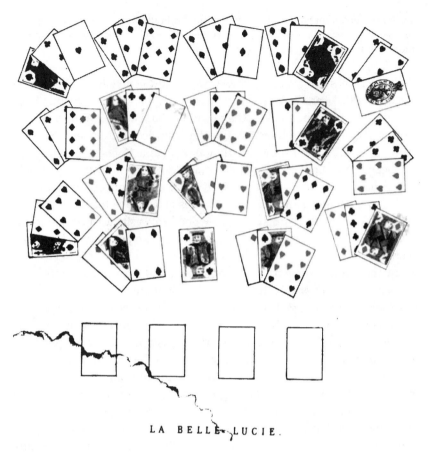

LA BELLE LUCIE.

Figure 3.4
"La Belle Lucie" Solitaire game (Cadogan 1876)

Miss Whitmore Jones would remark how times had changed and how Solitaire was *previously* considered a game for "idle ladies," but that in modern hectic times, the benefits of playing Solitaire had become widely appreciated:

In days gone by, before the world lived at the railway speed it is doing now, the game of Patience was looked upon with somewhat contemptuous toleration, as a harmless but dull amusement for idle ladies, and was ironically described as "a roundabout method of sorting the cards"; but it has gradually won for itself a higher place. For now, when the work, and still more the worries, of life have so enormously increased and multiplied, the value of a pursuit interesting enough to absorb the attention without unduly exciting the brain, and so giving the mind a rest, and, as it were, a breathing-space wherein to recruit its faculties, is becoming more and more recognised and appreciated.[16]

Solitaire games are generally single player, and generally entail taking a set of shuffled cards and placing them in order within the limitations imposed by the game rules. Solitaire games share a large number of conventions that enable players to transfer knowledge from one Solitaire game to another and for the description of a game to be quite terse. Lady Cadogan's 1876 book is introduced by an "explanation of terms" of Solitaire, illustrating how Solitaire is an open collection of mechanics of which every single game is a *selection*. This also means that once a player has played a few Solitaire games, new ones are easily learned due to the player's familiarity with the game mechanics. The popularity of Solitaire games on computers was furthered by such familiarity—players already know the original version played with cards, and a computer simply provides a convenient opportunity for playing Solitaire.[17]

Casual game design lowers the barriers to entry by requiring little knowledge of game conventions and small time investments, but the physical space required to play a game can also be a factor. Surprisingly, such barriers were considered an issue even in the early days of Solitaire games. A 1901 Solitaire collection praises the appearance of physically smaller card decks, hence lowering the barriers to entry to even the card version of the game: "Patience is now very generally played, as the one objection to it that used to exist—that it required so large a space to lay out the cards—has now been removed by the introduction of miniature packs, which have been specially made for it, so as to enable the most elaborate game to be displayed in the compass of about a foot square—a great boon to invalids confined to a couch, for they no longer require a table, but can set out these games on a tray, or even on a music-book."[18]

Solitaire games were from early on associated with specific audiences, with Whitmore Jones indicating that Solitaire was considered a ladies' game. In the history of games, this compares to the association of hardcore video games with young men and the association of contemporary casual games with a female audience seemingly similar to the perceived Solitaire audience. While the previous quote mentions the invalids benefitting from smaller decks of cards, the cover of the 1914 American edition of Cadogan's book[19] (figure 3.5) shows that a female audience is still associated with these games: a society woman with modern clothes, hairstyle and furniture is depicted enjoying a game of Solitaire in the evening.

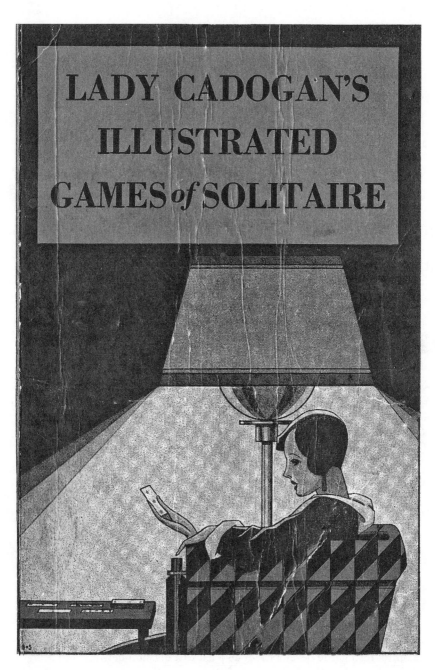

Figure 3.5
The 1914 edition of *Lady Cadogan's Illustrated Games of Solitaire or Patience* (Cadogan 1914)

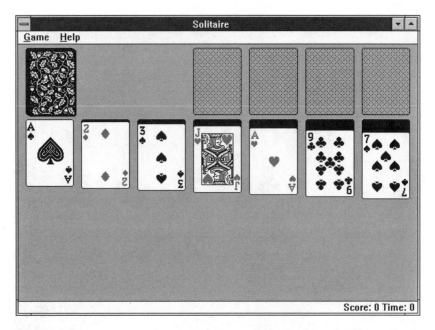

Figure 3.6
Solitaire for Windows 3.0 (Microsoft 1990, image courtesy of Rowan Lipkovits)

Solitaire was included as a standard application with Microsoft Windows 3.0 in 1990,[20] and hence made the leap to computers (figure 3.6). How popular is Solitaire played on computers? A Finnish study reports that Solitaire is the most popular *digital* game of both men and women, with 36 percent of women and 13 percent of men reporting it as their favorite.[21] Steve Meretzky has argued that the inclusion of Solitaire in Windows 3.0 was the beginning of casual games as such.[22]

Before discussing Solitaire further, I would like to compare the development of Solitaire to that of modern video games (such as casual games). Folk games like Solitaire were not designed by any one person, but developed slowly over time as players deliberately or by mistake introduced variations that they would then communicate to other players if they found the variation interesting. With the advent of commercial game development in the eighteenth century and later with video games, the design and design time of a game became tied to specific developers. Commercial game development also led to a division between the designers and players of a game. Around 1890, the U.S. game company Parker

Brothers (which would later publish *Monopoly*[23]) had experienced disappointing sales of a complicated strategy game called *Chivalry*, designed by George Parker. This led them to describe the following principles for future game development: "A customer—at any store, in any city—must be attracted by the intriguing name and colorful artwork on the cover of every Parker Brothers product. Each game must each have an exciting, relevant theme and be easy enough for most people to understand. Finally, each game should be so sturdy that it could be played time and again, without wearing out."[24]

These principles are surprisingly similar to the casual game design principles I have described: the fiction must be attractive; the game must be usable; the game must have a good visceral quality—juiciness. It is also notable how Parker's design principles introduce a division between the tastes of the developer and the tastes of the player: George Parker had been fond of his complicated strategy game, but realizing that this feeling was not shared by his audience, he decided to develop games for the perceived tastes of the audience *rather* than his own.

In many ways, video game development has spent decades catching up with the principles listed by the Parker Brothers. The history of video game development is partially a history of growing development teams and changing relations to the audience of games. Whereas early video games were often made by a single person, growing development teams led to an increased demand for game development to be properly planned, and for more consistent testing against a game's target audience. In a 2002 article, developers Mark Cerny and Michael John argue for a prototype-driven method for developing video games.[25] In this development method, a game is subjected to frequent *playtesting* with the target audience throughout development, but developers should *not* rely on focus group tests where audience are asked about their tastes, because this can only give information about "What's popular as of 10 minutes ago."[26] A more specific criticism against focus groups is that they "are poor at providing specific, actionable data that help game designers make their games better."[27] Commercial video game development often relies on observation of players playing the game in development, while some developers also doubt that the explicit opinions of players can be trusted.

The status of the audience in game development is thus much contended. A common complaint against the traditional video game industry is that developers are making games only "for themselves," with the casual games industry, like George Parker, proclaiming to make games for "everybody." A casual game developer describes his own position like this: "Hardcore developers make games for themselves ('I like that—let's put it in'), whereas casual developers make games for themselves and everybody else ('I like that, but let's make sure it works for my dad/sister/ receptionist too')."[28]

While this probably is an unfair generalization about the development of traditional hardcore video games, playing such games often requires game convention knowledge from other video games. To build a game on existing conventions is to run the risk of alienating an audience that does not know them. It could be predicted that casual games were therefore unlikely to build on previous games, but this is different in the two game types discussed in this book: while mimetic interface games borrow mostly from activities outside video game history (such as bowling or playing music), chapter 4 shows how downloadable casual games build heavily on the conventions of other downloadable casual games. This is probably due to the way such games are distributed: players have free access to trials of hundreds of similar games and therefore have time to absorb video game conventions. On the contrary, players of mimetic interface games are less likely to try more than a few games, as the games have to be purchased before they can be played.

All developers share a balancing act between making a game that conforms to tradition and making one that breaks new ground, between making a game for personal tastes and making one for an audience that the developer is not part of. Some of the developers interviewed in appendix C consider themselves to be part of their target audience, but one developer reports making games for the tastes of two fictional characters "Sophie and Marie."[29]

While developers make games for an audience that they may not be part of, the audience itself is constantly changing. Players learn new conventions and new skills during their lifetimes. Solitaire became a popular game on computers because players had become familiar with Solitaire conventions earlier in their lives (and because it was easily learned for people who were not familiar with the game). *Puzzle Quest* is arguably popular because it has dual entry points depending on a player's game knowledge. The negative user review of *Puzzle Quest* then illustrated how a player can be alienated from a game if he or she fails to understand the conventions it uses. During your lifetime, you collect knowledge from the games you play, and you use that knowledge for understanding new games.

All this leads up to the *game-playing time*. The game that successfully manages to get a player to start and keep playing adds to that player's knowledge of conventions. To play a new game is to learn new skills and conventions. The history of games leads up to your playing of an individual game; your playing of that game paves the way for playing future games.

Solitaire as a Proto-Casual Game

Solitaire only recently has become a game played on computers. Before a player begins a game of computer Solitaire, many things have already transpired. First there are thousands of years of game history, during which general game conventions have taken shape. Video games, as a subset, have emerged only during the last forty years. Then there is the time frame of designers, who develop a game in some weeks, months, or years, using their knowledge of previous games and their assumptions about the game's audience. And before an individual game of Solitaire is played in the present, the player has experience with other games and other media during his or her lifetime. These are the four time frames of games, which can be described as follows:

• *Historical time*   The evolution of games transcends the lifetime of any player or developer. Games and audiences evolve over millennia. Games and game genres are associated with specific audiences, but these associations also change over time. For example, it was once assumed that video games would only be played by young men, but this is changing with the rise of casual games. Furthermore, the introduction of Solitaire on the Windows operating system illustrates how a game can appear, become popular, and move between technological platforms because it is widely known and understood.

• *Design time*   Whereas traditional folk games were rarely, or only apocryphally, associated with specific designers, the rise of commercial board games and video games introduced a division between game designers and game players. Contemporary video game developers appear to be gaining a more nuanced view of the audience for their games, and casual game design is often framed as being more sensitive to audience demands than traditional hardcore game design. Casual game design introduces more distance between players and developers, with developers encouraged to create games for audiences other than themselves.

- *Player lifetime* Players see new games in the light of the previous games they have played. This game literacy is not a general game literacy but is tied to the specific genres a player has experienced. We identify new games based on the games and genres we already know.

- *Game-playing time* The preceding time frames lead up to the time of game-playing, where, hopefully, the player understands and enjoys the game.

Even before Solitaire became one of the most popular games played on computers, it matched casual game design principles very well: it is a usable game that players can play at their own pace, and supremely interruptible. Solitaire can be replayed indefinitely, so the game does not punish the player for failing by making him or her replay a level—the player is replaying the game anyway. Furthermore, the extended use of randomness relieves the player from some of the responsibility if a game is not solved. Finally, if the amount of space needed to place the cards of Solitaire have historically been a barrier to playing, players who own a computer can now play Solitaire without taking up any additional space. Computers just made Solitaire even more casual.

The successes of *Puzzle Quest* and computer-based Solitaire emphasize that games fail or succeed due to the *interaction* between game design and players. But even this is only a partial truth: in the bigger picture, neither game designers nor players start from scratch, but carry the history of games with them. Being aware of this is a requirement for understanding the casual revolution—and video games at all.

4 Innovations and Clones: The Gradual Evolution of Downloadable Casual Games

It is a feeling similar to playing Solitaire. You are totally relaxed, you cannot concentrate on anything else, but at the same time you can be thinking about other things in the back of your mind. I often play when I face a difficult problem. In my company I face various tasks that are hard to get started with. I already have the knowledge I need, so I play a game rather than go read a lot of books. Then the solutions come. It is like the game brings out a lot of tacit knowledge, as if the problem solving in the game maintains that skill, and that is a skill I need.
—A 55-year-old player of downloadable casual games[1]

Downloadable casual games are games in a specific *distribution channel*: while video games have traditionally been sold in stores, players of downloadable casual games go to a website such as RealArcade or Big Fish Games (figure 4.1) and download a free version of a game that can typically be played for sixty minutes (figure 4.2). After sixty minutes, the player has to buy the game in order to continue playing (figure 4.3). Developer Dave Rohrl describes the fact that players can try a game before playing as a way of selling games in an "open box."[2] This has some similarities with games like *Guitar Hero* or *Buzz*[3] discussed in chapter 5, where the shape of the controller gives players an idea of what the game is about before playing it. For any player unsure about which game to choose, this makes it much easier to make a purchasing decision.

As game genres contain specific conventions and mechanics, so does this distribution channel contain specific types of games. Downloadable casual games are almost exclusively two-dimensional games that take place in screen space. As of 2008, the most popular game type in this channel is hidden object games such as *Mystery Case Files: Huntsville*[4] (figure 4.4), followed by time management games (figure 4.5). In the early years of downloadable casual games, matching tile games was the bestselling

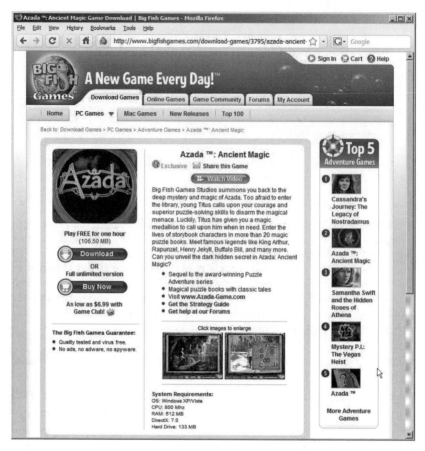

Figure 4.1
The Big Fish Games website

game type. In turn, the best-known matching tile game is the *Bejeweled* series from PopCap games, which has presently sold more than 25 million copies in different formats.[5] Figure 4.6 shows *Bejeweled 2 Deluxe.*[6]

Downloadable casual games played an important role in bringing industry and popular awareness to the fact that video games could reach outside their assumed audience of young men. A 2006 study of players of downloadable casual games reported that 71 percent of the audience was female,[7] with the majority being over 35 years of age. The study of casual players in appendix A shows an even greater skew: 93 percent of the respondents are female. This is worth noting as the game industry

Figure 4.2
Playing the downloaded game *Azada: Ancient Magic* (Big Fish Studios 2008)

has been reluctant to acknowledge the possibility of marketing outside the traditional market of young men. In the mid-1990s game developer Margaret Wallace worked at the company PF Magic making the virtual pet games *Dogz*[8] and *Catz*.[9] She told me the story of how difficult it was for the game industry to accept that they were making games that reached outside the traditional video game market:

When the company I was working at, PF Magic, was bought by Mindscape, we were integrated into a traditional console gaming company, including a company called SSI who made games like *Panzer General* and other war games, and Brøderbund who made *Prince of Persia*, and who had *Warhammer* and those titles. People at that company did not get us, they did not like us, they thought we were a joke. The traditional games industry did not know how to handle our games, *Dogz* and *Catz*, virtual pets. The sales people had to place our games in retail along with *Panzer General*. They did not know what to do with us.

Figure 4.3
Pay to continue playing

Figure 4.4
Mystery Case Files: Huntsville (Big Fish Studios 2005)

Figure 4.5
Diner Dash (Gamelab 2003)

Figure 4.6
Bejeweled 2 Deluxe (PopCap 2004)

We would get data from our customers saying that 14-year-old boys were not the dominant users of our games, it was girls and women. It was like a paradigm shift that people had the hardest time getting over. Even at PF Magic when we looked at the data, and saw that our users were more balanced in terms of the gender breakdown, people had the hardest time accepting that. We finally started talking about it and then we got acquired by Mindscape and then everything went bad. It was like an elitist club. People did not recognize that there was a wider player base out there.[10]

In the following section, I will examine the history of the game type that initially dominated the downloadable casual game channel: *matching tile games*. PopCap, developer of *Bejeweled* have claimed that its matching tile game started the phenomenon of casual games as such: "When we founded PopCap Games in 2000 and launched our first title, Bejeweled, we had little idea our modest jewel-swapping game would help to pave the way for a whole new genre of 'casual games.'"[11]

While this claim is subject to discussion, the history of matching tile games shows that the balance between innovation and cloning looks different from the perspectives of players and developers: developers tend to present a version of video game history that emphasizes *their* originality, explaining that *their* game is the original game that inspired other games (rather than the other way around). Players, on the other hand, have no reason to deny the connection between a new game and the games they have played before.

A Popular Mechanic with No Vocal Proponents

Matching tile games are quite simple: a large number of games can be described with very few parameters, not unlike Solitaire card games. By matching tile games I mean video games where the player manipulates tiles in order to make them disappear according to a matching criterion. In this chapter I will discuss matching tile games as being a *mechanic*, a typical set of actions for the player, rather than a genre. Matching tile games have been considered different things historically: at one point they were considered derivatives of *Tetris*; at another point a genre onto themselves; more recently, the matching tile mechanic is used as a minor mechanic in larger games. Additionally, matching tile games are interesting in that they may be one of the only game types with no vocal proponents, only critics. The *Puzzle Quest* review quoted in the beginning of chapter 3 encouraged players not to be "ashamed" of liking such games.

Where playing an imported Japanese game can be construed as a sign of game competence, matching tile games occupy perhaps the lowest rung on the cultural ladder, one of video game enthusiasts. Critics especially tend to complain of too many games in the subgenre of match-three games (usually referring to derivatives of *Bejeweled*): "On the big portals, at any hour, day or night, tens or hundreds of thousands of players gather to play Hearts, Spades, Canasta, chess, backgammon and a zillion shareware match-three games."[12]

PopCap, one of the leading developers and publishers of casual games, has this to say about matching tile games:

Q: *What kind of games [are] PopCap interested in publishing?*
A: Not just match-3 puzzle games! We're interested in pushing the boundaries of the casual games market with a variety of different projects.[13]

Some observers have expressed surprise at how long matching tile games remained popular in the downloadable casual games channel: "I used to preach that the world did not need another *match three bubble popper, Mahjong game*, or *card game*, but all of those game types have continued to sell in the Casual game space, and are even beginning to be considered genres."[14]

This low status of matching tile games may be a result of their low barrier to entry: these games are designed to be usable, and hence playing a matching tile game does not signal special knowledge of video games. This does not mean that we can declare matching tile games to be "bad" games, but in several ways they are at odds with a more traditional video game ethic that demand games to be challenging and punishing (this was discussed further in chapter 2).

A History of Matching Tile Games

Immature poets imitate; mature poets steal; bad poets deface what they take, and good poets make it into something better, or at least something different.
T. S. Eliot[15]

What would the history of a game or game mechanic look like? Some anthropological work has been done on game history: Stewart Culin's 1896 article on *Mancala, the National Game of Africa*[16] traces the spread of Mancala games geographically and historically, noting differences in rules and materials used to play. Writing the history of matching tiles games is slightly different in that the time span is much shorter in comparison to Mancala (twenty years rather than thousands of years). Matching tile games were developed mostly commercially and are generally attributable to individuals, as opposed to a folk game like Mancala that has no specific author. It is also not uncommon to see mostly journalistic histories of video game genres, such as real-time strategy games,[17] but I will give a more detailed account of how matching tile games have developed. It is not possible to include all matching tile games in this space, so I have selected games that have provided some type of innovation.

Figure 4.7 presents a family tree of the history of matching tile games, which I developed by examining as many games as possible, by reading

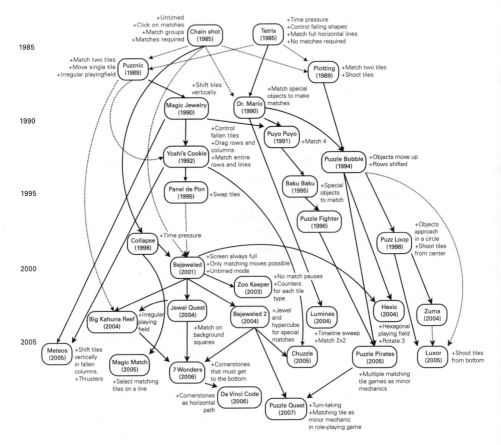

Figure 4.7
A family tree of matching tile games

developer interviews, and by soliciting comments for progressive versions of the history from developers and players. Arrows in the family tree indicate that the topmost game connected by an arrow in hindsight appears to have inspired the game below it. Except for a few cases I have not verified this, but the more speculative connections are indicated with dotted lines. As game players, it rarely matters to us whether actual inspiration took place: we may perceive a game as derived from another game, regardless of whether there is any truth to this. In other words, the family tree is a perception of the history of matching tile games.

From the top of the diagram, there are two progenitors of matching tile games, the 1985 *Chain Shot*,[18] also known as *Same Game* (figure 4.8) and the better known *Tetris*,[19] also from 1985 (figure 4.9). I cannot rule out the existence of earlier, little-known matching tile video games, but *Tetris* was an extremely successful game that spawned a number of imitators, and the influence of *Chain Shot* is visible at various points in the tree. Both games were originally noncommercial.

In retrospect, *Chain Shot* and *Tetris* foreshadow several trends in the following twenty years of matching tile game history. The two games diverge on four important counts:

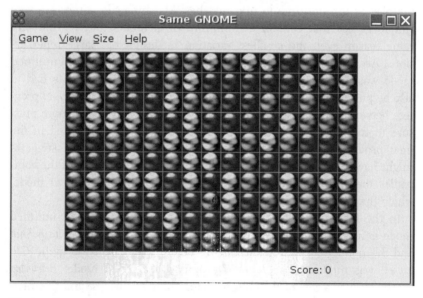

Figure 4.8
Same GNOME (Gnome Project 2006), a recent version of *Chain Shot!*

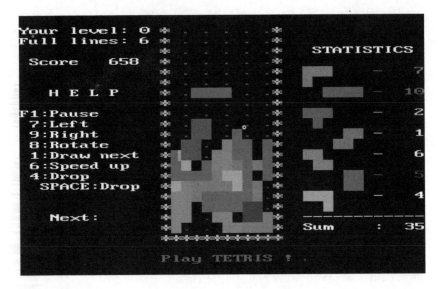

Your level: 0
Full lines: 6   STATISTICS

  Score   658

    H E L P

F1:Pause
  7:Left
  9:Right
  8:Rotate
  1:Draw next
  6:Speed up
  4:Drop
    SPACE:Drop

  Next: Sum   : 35

              Play TETRIS ?

Figure 4.9
Tetris (Pajitnov and Gerasimov 1985)

Time   *Tetris* puts the player under time pressure, but *Chain Shot* affords the players infinite time to find matches. Matching tile games are very simple games that contain a minimum of elements, but for the very same reason even the smallest variation in design has large repercussions. According to the *Bejeweled* developers, the inclusion of an untimed mode was quite controversial: "Numerous other enhancements had to also be put in place, like the inclusion of the meter that lets players progress between levels, and a timer that ticked down and added more pressure to the game. Of course the untimed version was included in the final product; something that [Bejeweled developer] Kapalka suggests might have been integral to its success.... He said that many of the companies they showed the game to were alarmed by the untimed mode, which they believed didn't require any skill to do well at."[20]

In the history of matching tile games, we can see that an untimed mode is not an entirely new development, but is a return to *Chain Shot* and the nondigital version of Solitaire before that. The 1998 game *Collapse*[21] was modeled on *Chain Shot*, but with a timed mode. *Bejeweled* can be seen as a mix of the obligatory matching of *Collapse* and the interaction of tile swapping in *Panel de Pon*[22] but with the untimed mode of the much earlier *Chain Shot*.

Figure 4.10
Dr. Mario (Nintendo 1990)

Manipulation   *Tetris* lets the player manipulate tiles *as they fall*, but *Chain Shot* lets the player manipulate tiles that *have* fallen. This difference divides games such as *Dr. Mario*[23] (figure 4.10), in which players control falling tiles, from games like *Chain Shot* or *Yoshi's Cookie*[24] (figure 4.11), in which the player manipulates tiles that have already *fallen*. The major subsequent innovation leading up to *Bejeweled* is the mechanic of swapping tiles as introduced in the 1995 *Panel de Pon*[25] (figure 4.12). Recent years have seen many variations of tile manipulation, with one of the more successful being the shooting of tiles by the player, originally found in the 1989 *Plotting*[26] (figure 4.13), but influencing most of the right side of the history tree up to the more recent games like the 2005 *Luxor*[27] and the 2004 *Zuma*.[28] Finally, the 2005 *Chuzzle Deluxe*[29] (figure 4.14) appears derived from *Yoshi's Cookie*, but features a constantly full screen— as in *Bejeweled*.

Match criteria   *Tetris* requires an entire horizontal line to match; *Chain Shot* requires the player to match tiles with similar colors. While *Tetris* has been hugely popular, its matching criteria of filling an entire row surprisingly has not been copied much in later games. Rather, all other

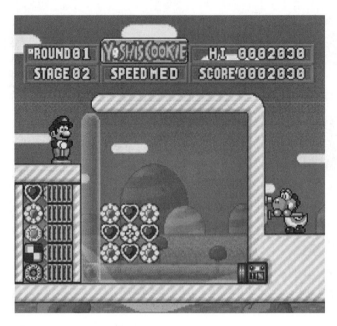

Figure 4.11
Yoshi's Cookie (Bulletproof Software 1992)

Figure 4.12
Panel de Pon (Intelligent Systems 1995)

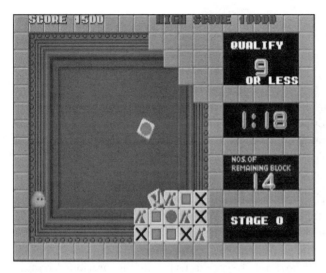

Figure 4.13
Plotting (Taito 1989)

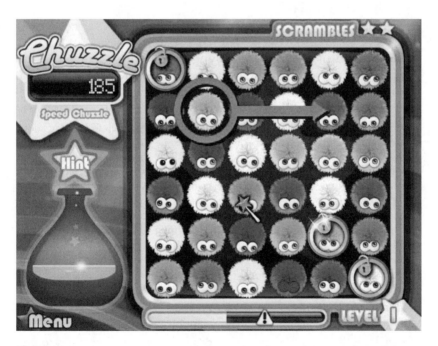

Figure 4.14
Chuzzle Deluxe (Raptisoft 2005)

games in the history here follow the *Chain Shot* model of having several types of tiles that can then be matched based on similarity.

Obligatory matches *Tetris* lets the player perform actions regardless of whether they lead to a match, but *Chain Shot* only lets the player perform actions that lead to a match. The overall effect of the latter is that the game requires fewer player actions and mouse clicks overall. In the family tree of matching games, most of the games to the left require the player to make matches, but most of the games to the right let the player perform other actions as well. The left side of the family tree is more strategic, and the right side of the family tree is more hectic.

From a design perspective, these four variations are quite special in that they are independent: the time mode of a game design can be modified without influencing the type of manipulation; the manipulation type can be modified without influencing the match criteria, and so on. Again, this is quite similar to Solitaire as discussed in the previous chapter; matching tile games are a well-defined design space that allows numerous variations based on a few simple building blocks.

Matching Tile Games and Developers

Jim Stern from iWin, makers of the 2004 *Jewel Quest*[30] (figure 4.15) has described their process of creating a new game by adding a small variation to *Bejeweled*:

Match-Three games have done well historically and have proven to be quite addictive. We wanted to take a familiar concept that people already enjoy and raise it to a level that is much more exciting and engaging than it's ever been.

With that in mind, we added new properties to the jewels (such as buried relics that require multiple matches before they can be removed and cursed items that can wreak havoc on your progress under special circumstances), new layouts (such as different shaped boards and areas that are inaccessible), and more importantly, a specific goal to complete each board (turning all the tiles to gold).

These relatively simple concepts, when combined in different ways, allow for great variation and ramping of play levels to provide hours and hours of challenging game play.[31]

This verifies that *Jewel Quest* was inspired by previous matching tile games, but it also explains the very gradual innovation in the family tree of matching tile games: every game adds only very small changes to previous games.[32] As Jim Stern states, this allows a game to capture an audi-

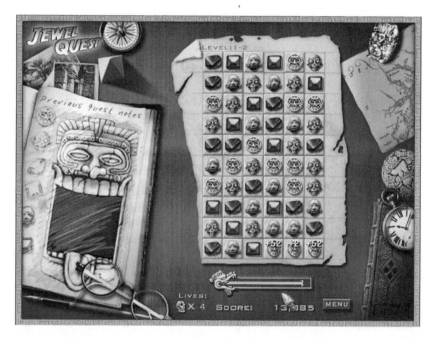

Figure 4.15
Jewel Quest (iWin 2004)

ence already familiar with the genre, while adding a few twists that give players new experiences.

The family tree shows how *Bejeweled 2* introduces special objects for big matches and how *Big Kahuna Reef,*[33] released later in 2004, adds an irregular playing field in addition to borrowing from *Jewel Quest*. All of these new elements are combined in the 2006 *7 Wonders of the Ancient World*[34] (figure 4.16) along with the introduction of special "cornerstones" that the player must move to the bottom of the screen.

Hence, *7 Wonders of the Ancient World* is a comparatively complex matching tile game that combines new features from at least three previous matching tile games. This does *not* mean that matching tile games are historically destined to become ever more complex, but it does foreshadow the changing status of matching tile games as will be discussed later.

Even more than other distribution channels, the casual downloadable game channel is characterized by the two opposing requirements of familiarity to the player and sufficient innovation to differentiate a game

Figure 4.16
7 Wonders of the Ancient World (Hot Lava Games 2006)

from other games on the market. This creates a somewhat schizophrenic environment of cutthroat competition between developers simultaneously trying to out-innovate and out-clone each other. Consider the contested bottom-right corner of the family tree in figure 4.7: much of the initial response to PopCap's 2004 hit *Zuma*[35] (figure 4.17) described PopCap as creators of an original game that subsequently had been imitated by others,[36] including the three 2005 games *Luxor*[37] (figure 4.18), *Tumblebugs*,[38] and *Atlantis*.[39] A 2005 interview with PopCap's director of business development emphasizes PopCap's prototype-oriented development method and mentions the large number of *Zuma* clones.[40] (The interview does not explicitly claim that *Zuma* was an original concept developed by PopCap.) Subsequently it became known that *Zuma* was in fact very similar to the 1998 arcade game *Puzz Loop*[41] (figure 4.19).

There have even been rumors of an impending lawsuit by *Puzz Loop* developer Mitchell against PopCap,[42] but the legal basis of such a lawsuit is not clear. Ironically, one reviewer described a new version of *Puzz Loop* as a "clone" of *Zuma*.[43] To further complicate the issue of which game

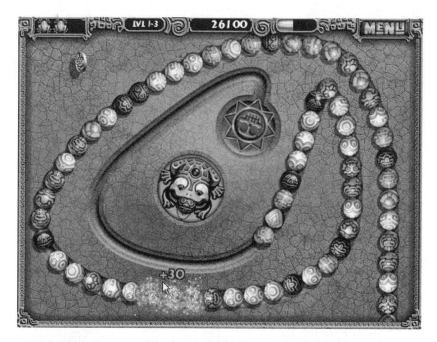

Figure 4.17
Zuma (PopCap Games 2004)

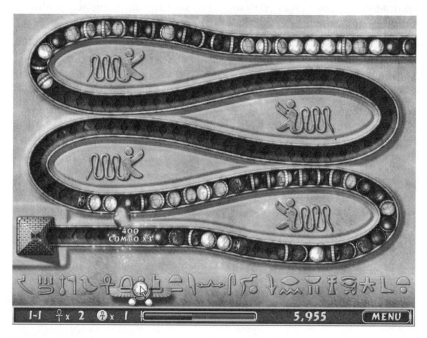

Figure 4.18
Luxor (Mumbo Jumbo 2005)

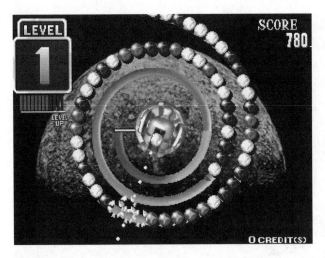

Figure 4.19
Puzz Loop (Mitchell 1998)

inspired which, Darren Walker of *Luxor* developer Mumbo Jumbo has in an interview downplayed inspiration from *Zuma* and emphasized the basic experimentation that lead to the game design: "When asked by the moderator about the influence of *Zuma* on *Luxor*, Walker hesitated, commenting: '*Zuma* was certainly a factor.' After thinking about how to integrate the basics of *Centipede* and *Galaga* with puzzle game mechanics, the developers worked from the core mission to have a game without negative in-game actions, such as anti-power-ups, that would discourage players."[44]

Yet, *Luxor* was received as a *Zuma* clone with minor innovations. It *is* possible to see a potential link between *Luxor* and the 1980 *Centipede*[45] (figure 4.20), but *Luxor's* similarities to *Puzz Loop* and *Zuma* are much more apparent.

As quoted, the developers of *Zuma*, *Puzz Loop*, and *Luxor* all exhibit a desire to be considered original and an anxiety about being seen as influenced by other games. Depending on which developer you ask, the history of matching tile games can be written three different ways, with *Zuma* as innovator, *Puzz Loop* as innovator, and *Luxor* as innovator (albeit inspired by other games) (figure 4.21).

Although the goal here is not to determine who actually inspired whom, I believe there are strong arguments for the type of history shown

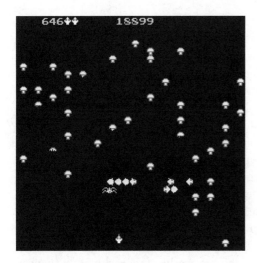

Figure 4.20
Centipede (Atari 1980)

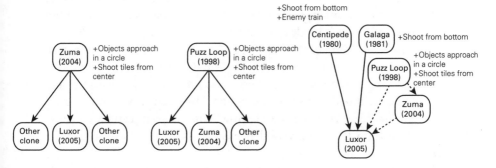

Figure 4.21
Zuma as innovator, *Puzz Loop* as innovator, *Luxor* as innovator

in the family tree in figure 4.7 with *Puzz Loop* inspiring *Zuma* and *Luxor*, and *Puzz Loop* being inspired by the earlier *Puzzle Bobble*[46] and *Plotting* in that it lets the player shoot tiles in order to make matches.

The Problem with Video Game History

This chapter traces a history of matching tile games during a period of more than twenty years. However, the history recorded here is not the only one that can be written, and it is quite selective: I have only focused on matching tile games and only on video games, thereby leaving out many sources of potential inspiration. The basic mechanic of matching can be attributed to nondigital games such as Mahjong solitaire and dominoes, and card games including Solitaire. Shooting tiles (such as in *Plotting*) seems derived from the Japanese game of Pachinko. Limiting the player to only making matching moves may have been inspired by the nondigital games of Peg Solitaire as well as Solitaire itself. Since there are a potentially unlimited number of external influences on matching tile games, limiting the focus to video games is a simplification that makes it possible to discuss the history of matching tile games at all.

In 1936, Alfred J. Barr created a diagram of the history of "Cubism and Abstract Art" for an exhibition at the Museum of Modern Art in New York (figure 4.22).[47] Edward Tufte points out that Barr's diagram only includes influences internal to the art world, and excludes influences from all other parts of society and history. Additionally, Tufte is critical of how all influences are mapped as unidirectional arrows, excluding mutual influences between artists or directions.[48] I think that this type of criticism does not render such a work of history impossible or false, but simply requires that it be clear about what it is a history *of*.

A more general criticism of this type of history is that it is a simplification of the way an art actually develops: Game creation and consumption are much more complex, and a huge amount of data is neglected and suppressed in order to reduce the relation between, say, *Yoshi's Cookie* and *Panel de Pon* to a single causal arrow. This criticism certainly is true, but only true in the same way that *any* theory is not the world, but a theory about the world.

While the history of matching tile games written here is not the only one possible, a history of matching tile games is not just a theoretical idea imposed upon the world. Rather, the knowledge of conventions— including those of game mechanics, genres, and interfaces—is an important aspect of the development and consumption of *all* games.

Are Matching Tile Games a Genre?

I have been identifying matching tile games as a game mechanic because these games historically have had an inconsistent status. For a long period of time, matching tile games were considered derivatives of *Tetris*, which was given the status of a prototype game. A 2001 review of a

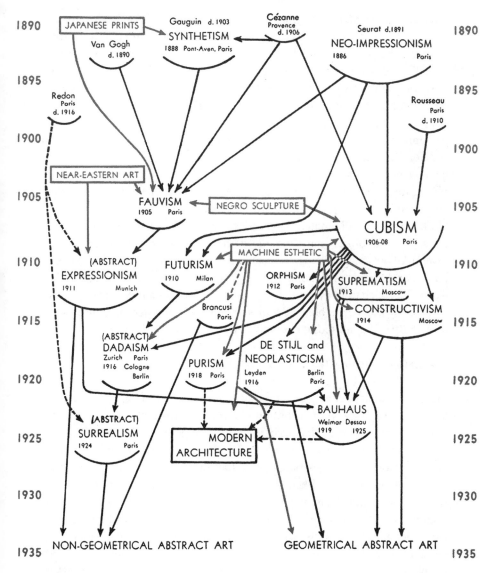

Figure 4.22
Cubism and Abstract Art (Barr 1966)

rerelease of *Dr. Mario* describes it as using elements from *Tetris*: "In the late 80s Nintendo had a great idea to make a puzzle game that borrowed elements from Tetris and Puyo Puyo and featured Mario's name in the title. That idea was realized in the form of the NES release, *Dr. Mario*."[49]

While the review describes *Dr. Mario* as borrowing elements from two other games, the review misses the fact that *Dr. Mario* was released *before* the game *Puyo Puyo* of which it is claimed to be derivative. Again, player perceptions of history can be quite subjective.

After the success of *Bejeweled*, "match-three" gained popularity as a genre label, but was generally invoked to mean a subset of "matching tile games." In one contemporary categorization, matching tile games are considered a *superset* of genres by distinguishing between "falling block puzzles," "tile-matching puzzles (Creation)" and "Puzz Loop variants."[50] In another, matching tile games are considered subgenres of several larger categories where match-three are part of "puzzle games," but "marble poppers" (*Zuma* and others) are part of "action & arcade."[51] Genre labels change over time, but as with *Tetris*, a single game can be sufficiently popular to become a prototype name for a genre, and the success of *Bejeweled* led to the creation of the "match-three" label.

The End of Matching Tile Games?

Matching tile games are not as popular as they used to be. In the downloadable casual games distribution channel, a survey of sales data from 2001 to 2007 showed that matching tile games were the second most popular game type in 2004 and by far the most popular in 2005, but only the fourth most popular of a handful of genres in 2006 and 2007.[52] A publisher I talked to in 2008 asserted that match-three now was a niche genre. Matching tile games were first eclipsed by time management games led by *Diner Dash*[53] (figure 4.5) and later by hidden object games led by *Mystery Case Files*[54] (figure 4.4). Is this the end of matching tile games? Not at all, for matching tile games are not going away, but simply moving from the high-profile status of a game type to the lesser role of a minor mechanic within larger games. *Puzzle Quest*, for example, uses matching tile games as one mechanic in combination with role-playing game mechanics. Matching tile games are also acquiring a different status today, that of a "traditional" game, in other distribution channels. *Bejeweled* and *Zuma*, for example, are now sold on cell phones and MP3 players, and are playable as in-flight games on airplanes along with Solitaire games. Why? Because these games, like Solitaire, are now so well known that they can be assumed to be immediately playable by a large part of the population that happens to buy MP3 players or travel by air. It is not that matching tile games are going away, but rather they have moved from the top tier of downloadable casual games into other contexts and onto new platforms.

A Channel of Moderate Innovation

The mechanic of matching similar items is not new in game history: it can be found in a wide range of games including Solitaire, dominoes, and Mahjong. As in the discussion of Solitaire card games, matching tile games illustrate how a game design is not tied to a specific technological platform, and how usability and low barriers to entry are important issues for *any* game.

Not everybody is happy with downloadable casual games. Game developer Eric Zimmerman told me the story of his hope for, and later disappointment with, this distribution channel: "There was an idea that downloadable games could be a renaissance for innovation in terms of theme, content, and gameplay on a smaller scale in terms of budget size and production scope," he said. "But in fact, the downloadable casual games industry has evolved into something *more* clone-driven and genre-bound than the so-called hardcore game industry that it sought to make and end-run around. So, the downloadable casual games industry has become a parody of itself."[55]

This is *developer* disappointment with the downloadable casual games channel. From a *player* perspective, things may look entirely different. The simplicity of the games in the downloadable casual games distribution channel makes the inherent conflict between innovation and cloning appear a little sharper than elsewhere. Players have different interests in and amounts of tolerance for experimentation and variation. If we are looking for a relaxed fifteen-minute experience, we often *do* want some variation, but a radically innovative game will in all likelihood not work. The downloadable casual games channel may not be a place for radical game design innovation, but then it *does not need to be*: downloadable casual games compete not with experimental games or modern art, but with Solitaire games and crossword puzzles. It is the distribution channel for absorbing games with moderate innovation.

5 Return to Player Space: The Success of Mimetic Interface Games

We don't play *Parcheesi/Sorry!* with the kids, because it is too complicated for them—they are only three and a half years old. With the Wii, on the other hand, the way that you do something and see a reaction on the screen, the way you tilt the controller and see something on the screen—that is something different. You cannot give them PlayStation controllers; those are a little too advanced with too many buttons. With the Wii, we can see on the kids that it just works for them, they can use that immediately.

We play the Wii with friends, at social events. We have also played it with the in-laws who are both around sixty. They play it eagerly, and they ask if we shouldn't play the game one more time.
—Interview with a Wii-playing family[1]

Here is the formula for the success of the Nintendo Wii and games like *Guitar Hero* and *Rock Band* shown in figures 5.1–5.4: they have physical interfaces that mimic the action in the games. The *Wii Sports* tennis player must swing his or her arm in order to swing the racket in the game; the *Guitar Hero* player must "strum" the plastic guitar; the *Wii Fit*[2] player must actually strike a yoga pose.

This makes mimetic interface games easier to learn than traditional video games, and it adds new types of fun—easier, because players can use their preconceptions of tennis, singing, or playing guitar in order to play the game; fun in new ways, because players can learn from watching each other, because failure becomes an enjoyable spectacle, and because the games thereby become more immediately social than those played with standard game controllers.

Where traditional hardcore games focus on creating worlds, on *3-D space*, and downloadable casual games focus on the experience of manipulating tangible objects on *screen space*, mimetic interface games emphasize the events in *player space*. Mimetic interface games encourage us to

Figure 5.1
Wii Sports bowling player (Noel Vasquez/Getty Images)

Figure 5.2
Guitar Hero World Tour players (AP/Wide World Photos/Damian Dovarganes)

Figure 5.3
Guitar Hero player (image by Michael McElroy)

Figure 5.4
Game designer Shigeru Miyamoto plays *Wii Fit* (AP/Wide World Photos/Tine Fineberg)

Console Standards

Tandy
1 Stick
1 Button

Atari 2600
1 Stick
1 Button

ColecoVision
1 Stick
2 Buttons
1 Number Pad

Atari 5200
1 Stick
4 Buttons
3 Options
1 Number Pad

NES
1 D-Pad
2 Buttons
2 Options

Sega Master System
1 D-Pad
2 Buttons

Genesis
1 D-Pad
3 Buttons
1 Option

SNES
1 D-Pad
4 Buttons
2 Shoulders
2 Options

Sega CD
1 D-Pad
6 Buttons
2 Options

N64
1 D-Pad
1 Stick
6 Buttons
3 Shoulders
1 Option

Dreamcast
1 D-Pad
1 Stick
4 Buttons
2 Shoulders
1 Option

Playstation 2
1 D-Pad
2 Sticks
4 Buttons
4 Shoulders
3 Options

Gamecube
1 D-Pad
2 Sticks
4 Buttons
3 Shoulders
1 Option

X-Box Old
1 D-Pad
2 Sticks
6 Buttons
2 Shoulders
2 Options

X-Box New
1 D-Pad
2 Sticks
6 Buttons
2 Shoulders
2 Options

X-Box 360
1 D-Pad   1 Stick
2 Sticks   2 Shoulders
6 Buttons   Motion
2 Shoulders   sensitive
2 Options   Aimed

Wii Mote
1 D-Pad
3 Buttons
1 Shoulder
4 Options
Motion
sensitive
Aimed

Wii Arcade
1 D-Pad
2 Sticks
4 Buttons
4 Shoulders
3 Options

Specialty Controllers

Atari Tennis
1 Knob
1 Button

ColecoVision
1 Stick
4 Buttons
1 Number Pad

NES Light Gun
1 Button
Aimed

NES Power Glove
1 D-Pad
2 Buttons
15 Options
Motion Sensitive
Aimed

Dreamcast Fission
1 Stick
4 Buttons
1 Reel
Motion Sensitive

Keyboard and Mouse
110 Buttons
2 Number Pads

2 Buttons
1 Scroll Wheel
Motion Sensitive

SNES Super Scope
1 Button
2 Options
Aimed

PS2 Guitar
5 Buttons
2 Options
1 Strum
1 Wammy
Motion Sensitive

Portable Systems

Game Boy	Lynx	Game Gear	Virtual Boy	Game Boy Color
1 D-Pad	1 D-Pad	1 D-Pad	2 D-Pads	1 D-Pad
2 Buttons	4 Buttons	2 Buttons	2 Buttons	2 Buttons
2 Options	5 Options	1 Option	2 Shoulders	2 Options
			2 Options	

GBA	GBA SP	DS	Game Boy Micro	PSP	DS Lite
1 D-Pad	1 D-Pad	1 D-Pad	1 D-Pad	1 D-Pad	1 D-Pad
2 Buttons	2 Buttons	4 Buttons	2 Buttons	4 Buttons	4 Buttons
2 Shoulders	2 Shoulders	2 Shoulders	2 Shoulders	2 Shoulders	2 Shoulders
2 Options	3 Options	1 Touch Screen	2 Options	6 Options	1 Touch Screen
		2 Options			2 Options
		1 Stylus			1 Stylus

Figure 5.5
A history of game controllers (image courtesy of Damien Lopez)

imagine that the game guitar is an actual guitar that we play on, and the Wii controller is an actual tennis racquet we swing to hit the ball. Where more traditional three-dimensional games force players to imagine a bodily presence *in* the game world, mimetic interface games allow players to play from the perspective of their physical presence in the real world. While many of the popular mimetic interface games *do* have three-dimensional worlds, those worlds can often be ignored during game play, as in the case of *Guitar Hero*. In the case of the *Wii Sports* games, the controllers support the illusion that the player space is continuous with the 3-D space of the game, that the two types of space are one.

In a historical perspective, mimetic interfaces, like casual games as such, are partially a return to the early days of video games. Early arcade games had no standardized set of controls, so individual games often had custom mimetic interfaces. For example, the 1975 *Destruction Derby*[3] featured steering wheels, and the 1977 submarine game *Sea Wolf*[4] had a periscope. The mimetic interface all but disappeared with the standardization of the game controller from the mid-1970s onward, as illustrated in figure 5.5. The illustration also shows how controllers historically have become increasingly complex, adding ever more buttons, sticks, and directional pads. This is a common gripe with video game controllers: one

Figure 5.6
Virtua Tennis 3 (Sumo Digital 2007)

player complains that "these systems look like Mission Control for NASA, so I never play with them. I can't. There are too many buttons."[5] Mimetic interfaces are a backlash against these complex and counterintuitive game controllers. Finally, figure 5.5 also shows how home consoles *have* had specialized mimetic interface controllers like light guns and fishing reels, but that the Wii controller is the first *generalized* mimetic interface: it can be used for a variety of games.[6]

How Mimetic Interface Games Work

The difference between traditional video game interfaces and mimetic interface games can be simply illustrated with the manuals from two tennis games, *Virtua Tennis 3*[7] and *Wii Sports* tennis shown in figures 5.6–5.9. Where *Virtua Tennis 3* textually describes the role of each individual button on the controller, *Wii Sports* can illustrate the physical movements of the *body* of the player. In *Virtua Tennis 3*, there is an arbitrary relation

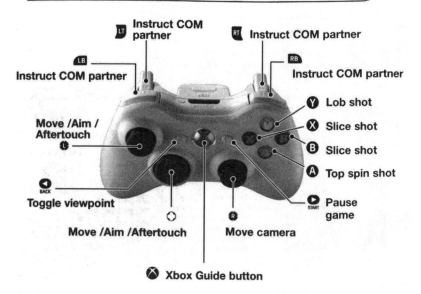

CONTROLS

Instruct COM partner — LT

Instruct COM partner — RT

Instruct COM partner — LB

Instruct COM partner — RB

Move /Aim / Aftertouch

Y Lob shot

X Slice shot

B Slice shot

A Top spin shot

Toggle viewpoint — BACK

Move /Aim /Aftertouch

Move camera — R

START Pause game

Xbox Guide button

Figure 5.7
Controller instructions for *Virtua Tennis 3* (Sumo Digital 2007)

Figure 5.8
Wii Sports tennis (Nintendo 2006)

Simply swing the Wii Remote to play a match of doubles tennis. You don't need to press any buttons. Each player needs his own Wii Remote.

The team that wins the required number of matches first wins the game.

 Holding the Wii Remote *(for right-handed players)*
- Hold it as would hold a tennis racket.
- Put your wrist through the Wii Remote strap and fasten it to prevent it from escaping your hand.
- Swing gently, and do not let go of the Wii Remote.

Serving

- Swing the Wii Remote up...
- and then down to serve the ball.
- Toss the ball up. You also can toss with the A Button.
- Hit the ball.

Controlling the Ball

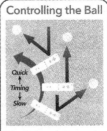

The timing of your stroke determines the direction the ball travels.

Strokes

Forehand   Backhand

- Judge the ball direction and swing accordingly. When the ball comes to your right, swing on the right side. When the ball approaches your left side, swing on the left. You can lob and give the ball a spin depending on how you hit.

Figure 5.9
Controller instructions for *Wii Sports* tennis (Nintendo 2006)

between the buttons pressed to hit the ball and the action in the game. In *Wii Sports*, there is a basic similarity between the movements of the player and those of the player's character in the game.

In chapter 2 I introduced Michel Beaudouin-Lafon's terms to describe how a game lets a player operate on domain objects by way of some interaction instruments.[8] In the case of *Wii Sports* tennis, a player uses the Wii remote, the interaction instrument, to control a domain object, the character in the game. *Virtua Tennis 3* has the same type of domain object, but the interaction instrument is the standard Xbox 360 controller. In Beaudouin-Lafon's terms, the Wii game has high compatibility because the movements of the player and the movement of the player-character are quite similar, but the interface of *Virtua Tennis 3* has low compatibility.

The high degree of compatibility of mimetic interface games has several implications for how these games can be used. Consider the *Guitar Hero III*[9] box shown in figure 5.10. A player does not need to have played guitar to understand this game, but will have been exposed to the general posture of a guitar playing through media or elsewhere. The player knows that the left hand generally moves up and down the fret board, while the

Figure 5.10
Guitar Hero III guitar and packaging (image courtesy of Activision)

Figure 5.11
Guitar Hero III screenshot (image courtesy of Activision)

Figure 5.12
Buzz!: Quiz TV (Relentless Software Ltd. 2008), image courtesy of Sony Computer Entertainment Europe

right hand generally strums near the saddle of the guitar. Furthermore, the game also draws on common cultural representations of guitar playing in rock music (figure 5.11). Presented with a physical guitar controller, a player thereby starts with not only a good idea of the general physical activity in the game, but is also presented with an activity that is commonly culturally represented as positive.

Take *Buzz!*, a series of quiz games especially popular in the European market (figure 5.12). The game comes not only with a game disc, but also with four custom quiz game controllers with colored buttons (figure 5.13). To anyone who has ever watched a game show on TV (i.e., most people), the combination of the game show host on the packaging and the controllers openly signals that this is a quiz game in which players are supposed to press the buttons to answer questions. In fact, the developer of the game, Relentless Software, put special effort into making sure the game was packaged in a *transparent* box. This makes it obvious for potential buyers what the game is about if they see it on display in a supermarket or a game store.[10]

Magic Crayons

"I'm not great at it," guitar player Slash said. "And a lot of that has to do with the fact that it's hard for me to get rid of 30 years—whatever it is—

Figure 5.13
Buzz! The Mega Quiz box (Relentless Software Ltd. 2007)

20-some-odd years of playing in a certain way and then all of the sudden become accustomed to pressing some buttons and stuff. I have these little things that I'm so used to doing that when I'm playing 'Guitar Hero' it sort of screws me up."[11]

Guitarist Slash of Guns N' Roses fame confesses that although he features on the cover of *Guitar Hero III* (figure 5.10), his real guitar skills hamper his *Guitar Hero III* playing. This is the other side of mimetic interface games: *Guitar Hero* is different from playing guitar; playing *Wii Sports* tennis is not exactly the same as playing physical tennis. Having played Wii tennis for more than a few minutes, most players realize that many movements that look like the swinging of a racquet do not actually move the racquet as you'd expect. Much energy can be saved by replacing the large arm movements of real tennis with small flicks of the wrist. The controls of Wii tennis do have a higher degree of compatibility than *Virtua Tennis 3*, while at the same time *Virtua Tennis 3* allows for greater control over the character in the game. In that way, while mimetic interface games on the surface seem quite similar to the activity they represent, they also offer barriers to expert guitarists, tennis players, and so on. Game designer Chaim Gingold has coined the term *magic crayon*[12] to describe how taking away possibilities from the player can make it more likely that the player will produce something pleasing. Mimetic interface games are generally such magic crayons: they make it easy for players to experience competence—to play tennis well, to complete a rock song, to perform a choreographed dance.

Taking possibilities away from players is not a problem because most video game players, of any age, are neither expert guitarists nor top tennis players. One senior center in Medford, Massachusetts, advertised its new Wii system like this:

Stay cool: Come to the Senior Center!
Wii games, while not as strenuous as the real thing, do promote some physical activity: players need to use their arms and legs, and while doing so receive a moderate workout! The virtual game[s] are very realistic, and you don't have to worry about lugging around a bowling ball or golf clubs.

Come and play on our large screen and get that old rush from playing your favorite sport![13]

Here, the Wii is described as attractive due both to its similiarity to the real sports ("get that old rush"), and to its difference ("not as strenuous as the real thing"). This duality of similarity and difference is common to *all* nonabstract games—all such games have a level of abstraction or simplification of what they represent:[14] you may play a racing game in which cars can never run out of gas, a sports game where you can never be injured, or even a guitar game in which the guitar never needs tuning and the strings never break. *Wii Sports* tennis will infuriate a top tennis player with the simplicity of its controls, but it gives a non-player of tennis the experience of being skilled. While swinging my arm to swing a tennis racquet has a high degree of compatibility, the characters in *Wii Sports* tennis move by themselves, without my input; where regular tennis allows me to control ball direction by turning my body, rotating the tennis racquet, and so on, this has been reduced to a question of timing in *Wii Sports* tennis. *Guitar Hero* is similar in this respect: while the general pose of the player (left hand on frets, right hand strumming) has a high degree of compatibility with the actual playing of guitar, the left hand must press the five colored individual buttons rather than press the six strings of a guitar at the various fret positions. A trained guitar player will experience the lack of fine-grained control over the music as severely limiting, but the majority of players will experience the same limitation as extremely empowering, allowing them to perform a well-known tune that they would be unable to play on a real guitar. Mimetic games are magic crayons, taking possibilities from players in order to give a feeling of competence.

This magic crayon quality leads to the obvious objection against *Guitar Hero* and *Rock Band*, which is that as a player you are not *really* playing guitar, but only pretending. On some level it is true that these are not real instruments, but what makes them not real? The basic experience of playing these games is that if you do not press the buttons correctly there is no music, but if you press the buttons correctly, music appears—it feels as if you *are* making music. Interestingly, this is quite similar to learning to play an instrument: playing sheet music on the piano at first feels exactly like playing *Guitar Hero*: you follow a notation telling you what to do. If you press the right keys on the piano, music appears. If you press the wrong keys, music does not appear. It is only through practice that you begin to feel a direct connection between the piano keys and the notes that come out of the piano. Subjectively, playing *Guitar Hero* isn't any less real than playing a piano is when you first begin to learn to play, and this is probably why you *do* feel as if you are playing music, playing *Guitar Hero*.

At the same time, music games show big differences between instruments: singing or drumming in *Rock Band* or playing the congas and maracas games of *Donkey Konga*[15] and *Samba de Amigo*[16] does prepare you for "actual" singing and drumming; being able to sing and drum does make it easier to sing and drum in the games. The simple reason behind this may be that guitars are more complicated and have steeper initial learning curves than drumming or singing. A guitar game therefore needs to be more of a magic crayon in order to reach non-guitar players.

The Space of Mimetic Interface Games

When you wave the Wii remotes into the air as if you were boxing, then the whole party laughs, even though only two people are playing.
—Interview with a Wii-playing family[17]

Q: Is there a game design art to maximizing the value of existing social relations between players?
A: For a while in our prototyping of *Rock Band* there was this feeling that when you were playing, everybody was playing their individual part. Even if you were four people standing next to each other, the amount of interaction you had with the other players was almost nil. We did a lot of subtle things from prototype to prototype to make it so you actually did feel a visceral connection to the other people in the room, so you weren't just playing your part; you were also playing with other people in the room. That wasn't something that was immediately apparent. That took a lot of work. One thing was the way in which you save your band mates, one is the overall design of the user interface, to focus it so you knew what other people were doing as well.
—Interview with Harmonix, developer of *Rock Band*[18]

Mimetic interface games are easy to learn because they draw upon familiar conventions from outside video games, but the large-scale physical movements that players perform also make it easier to learn to play *by looking at other players*. It is much simpler to learn by watching large physical movements than by watching someone using a traditional video game controller, and this factor increases the social value of these games. The Wii-playing family previously quoted explains how simply looking at other players is enjoyable, and Nicole Lazzaro has noted how mistakes are considered comic when a game is played in a group.[19] Chapter 6 shows how mimetic interface games are also a return to the social dynamics of traditional board and card games, in part because the space of mimetic interface games is similar to such traditional games.

I claim that mimetic interface games shift focus from the three-dimensional space created by the game graphics, to the concrete player space. This may sound surprising given that many of the games discussed, such as *Rock Band* and *Guitar Hero* (figure 5.11) actually do present a 3-D space. The screenshots show an elaborate world, and a camera pans and zooms following conventions from rock videos. Yet *all the gameplay-relevant information stands still* and is not influenced by the panning of the camera—even though much effort has been put into presenting in-game characters of a band playing. All information that is directly relevant for game-playing is anchored to the screen space rather than to the 3-D space. Subjectively, I find that I *do not see* the band while playing *Rock Band*, since the images of the band and the crowd are not immediately relevant to my playing. Their primary function is to add mood, to add a spectacle for those not playing, and to add juiciness: positive feedback when the player does well. This also means that even though the game presents a three-dimensional space, players do not have to imagine a bodily presence in that space.

The interest in player space versus 3-D space can be gauged via the popular photo-sharing service Flickr.com: of the top twenty pictures named "Wii tennis," seventeen show people playing the game; of the top twenty pictures named "Gears of War," only *one* shows a person playing the game.[20] For mimetic interface games, players are much more likely to upload photos of *players* than of games—the events in player space are what players find interesting and memorable in these games. In fact, the advertising for the Nintendo Wii console promoted the spectacle of player space very openly by showing pictures of happy game players across age and gender rather than of the games themselves.[21]

The Rise of the Minigame

Some of the most popular Nintendo Wii titles, such as *Wii Sports* and *Wii Play*[22] are not single games, but collections of a number of short so-called *minigames*. While *Rayman Raving Rabbids*[23] does have a single overarching game that you can complete, one of the motivations for playing that game is that it makes available for play various zany multiplayer games such as cow-throwing (figure 5.14). In the introduction I told the story of

Figure 5.14
Rayman Raving Rabbids cow-throwing minigame (Ubisoft Montpellier 2006)

how many people in late 2006 were surprised to see their friends and family suddenly take an interest in playing video games—the games they were playing were mostly minigames like these.

The *Guitar Hero* and *Rock Band* series have the same duality as *Rayman Raving Rabbids*. For example, *Rock Band 2* lets the player complete the game in a *world tour* mode by playing songs in a number of cities around the world, but players are also free to simply play one of the available songs instead. The rise of the minigame and the short-session multiplayer game brings another layer of flexibility in game design, another way of letting players use a game as they wish.

The Success of Mimetic Interfaces

To summarize, mimetic interface games reach a broad audience by engaging players in culturally well-known and valued activities, such as tennis or guitar playing, via interfaces that mimic the physical actions of these activities, allowing players to more quickly exhibit skill in the video game version of the acitivity.[24] I do not mean to say that mimetic interfaces are the *best* way to create games, but they work well for creating social games that are easy to learn. Mimetic interface games return to the type of social interaction found in traditional board games, card games, and party games, giving players a face-to-face experience, even with a video game.

The success of mimetic interface games, then, is about four key qualities:

• Improved usability because the mimetic interface makes it easy for players to use their preexisting knowledge of the real-life activity to play the game;
• Ease of use, since mimetic interface games generally are easier to play than the activity they represent;
• Shift of focus to the player space in which the game is played; and
• Shift of focus to the existing social relations in that space.

Mimetic interface games have revolutionized video games by borrowing liberally from outside video games, and by reaching back to early, non-digital game forms.

6 Social Meaning and Social Goals

Q: Do you think about how to maximize the value of the social situation the game is played in?
A: Always. That has always been the key to *Buzz!*, what we call off-screen interaction. The chance of my video game being able to entertain you and make you laugh more than your friends sitting next to you, is slim to none. I am not going to be able to do it because I am not there; I am just a fairly simple computer program that reacts to a set of statistics.

It is all about, *"How come you knew that question about Van Halen? I didn't know you were a Van Halen fan? Oh, that was when you grew your hair long when you were sixteen, oh I remember, that was bad, you were going out with so-and-so."* I can't ever hope to replicate that, but hopefully the people on your sofa can. That is the idea, to try to bring those kinds of things around.
—Interview with David Amor, developer of the quiz game *Buzz!*[1]

In this quote, game developer David Amor describes off-screen interaction as the key to social games: social game design isn't about creating a game that is strategically *deep* as much as it is about making sure that the game, in turn, creates interesting interaction between players. *Any* multiplayer game takes on meaning from the social relations between players, but mimetic interface games such as *Buzz!* (figure 6.1) are close to traditional board or card games because much of their action happens off-screen; because they encourage interaction in player space. To better understand the success of mimetic interface games, this chapter looks both at multiplayer video games and traditional board games, to show how multiplayer games take on meaning from where and with whom we play them. Game designer Richard Garfield notes that "a particular game, played with the exact same rules will mean different things to different people," and he uses the term *metagame* to describe these

Figure 6.1
Buzz! Players (image courtesy of Chris Hinkley)

differences.[2] Here, I will focus specifically on one aspect of the meta-game: the social relations between players.

In the introduction I discussed the *pull*, the subjective experience of looking at a video game and wanting to play it. In a multiplayer game, the pull is closely related to who you are playing with; we are more attracted to playing with some people than with others. Who you play with also influences how you play: while we can probably agree that we want to win when playing games, there is more to it than that. If you are far ahead in a game, will you play a little badly in order to make the contest more even? Playing against your boss, your partner, or a child will most likely make you pause to consider, for example, will the child cry if he or she plays badly and loses? Will a defeated boss hold a grudge? And if you play against someone with whom you wish to be romantically involved, does that make you more or less likely to compete fully against this love interest in the game? This is the ambiguity of multiplayer games: they simultaneously carry a notion of a noble contest *and* have a wide range of undetermined social consequences, of meanings.

Figure 6.2
Animal Crossing: arriving in a new town and buying a house (Nintendo EAD 2002)

Take *Animal Crossing*[3] (figure 6.2), which begins with the player arriving in a new town, buying a home that is more expensive than the player can afford, and then becoming integrated in the social fabric of the town by working off the debt toward the character that sold the house. Over time, players can decorate their houses and get to know the in-game characters. More innovatively, the game synchronizes itself with the clock in the GameCube console, and characters will complain if you have not played for a while. Furthermore, although only one player can play the game at a time, it allows members of a family (for example) to play at different times but then to send messages and gifts to each other, which are received the next time a recipient plays (figure 6.3). How emotional and socially meaningful can it be to send messages and gifts to other people within such a game? Take the short Korean comic *Animal Crossing Is Tragic*[4] (see figure 6.4) that became popular on Western blogs in 2007.

This comic strip tells a very emotional story about the narrator's actual experience of playing *Animal Crossing*, in which he received in-game messages and gifts from his recently diseased mother. I will attempt to explain why so much emotion and meaning can take place in a game. First, where do these emotions come from? The events and emotions recounted by the comic were clearly not included *in* the game box. Rather, the narrator's playing of *Animal Crossing* derived meaning from the social

Figure 6.3
Animal Crossing: mailing a letter to another player

context in which he played the game. What, then, is it about *Animal Crossing* that makes it possible to attach so much social meaning to the playing of the game? The events retold in the cartoon are possible because *Animal Crossing* allows players to send messages and gifts to other players, and thereby perform socially meaningful actions. In other words, the game's messaging and gift-giving functions makes the game take on meaning from the context in which it is played. It is then because the game allows meaningful asynchronous gestures that the mother can perform friendly gestures toward her son, from beyond the grave, so to speak. The characters who ask about the mother's whereabouts then just add an extra layer of tragedy. *Animal Crossing* is a social game because it has many points onto which preexisting social meaning can be attached.

To take a less dramatic example, *Parcheesi* (or *Sorry!* or *Ludo*) is a strategically shallow abstract game in which chance plays a large role. *Parcheesi* is also a well-known social game. Why is it social? As always, describing games as sets of rules makes them sound rigid and unemotional. Finnish game researcher Aki Järvinen has explained that the reason we are often emotional when playing simple games is that pursuing goals is very emotional for us in general.[5] Our pursuit of game goals makes the playing of games emotional even if we cannot point to any emotional content in the rules of a specific game. I think this can be extended to multiplayer games by thinking about how goals work in so-

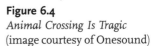

Figure 6.4
Animal Crossing Is Tragic
(image courtesy of Onesound)

cial contexts: players play for personal goals, are aware of the goals of other players, and the shared understanding of intentionality makes game actions socially meaningful. The *Animal Crossing* example is tragic in part because the narrator is aware that his mother was aware of his goals in the game. The shared understanding of goals allows helpful and not-so-helpful in-game actions. If players did not share this mutual understanding of goals, in-game actions would not have the same social and emotional value.

Why is the strategically shallow game of *Parcheesi* a social and emotional game? Because the capturing mechanic allows you to perform actions that are unhelpful (if you capture) or helpful (if you refrain from capturing), and because every choice about whether to capture must take into account three considerations: first, you want to win; it is usually a good idea to capture a piece. Second, you want to balance the game; if someone is far behind, it makes the game somewhat less fun if you capture. Third, you want to manage the social situation; depending on whom you are playing against, both capturing and refraining from capturing may have social consequences you do not wish for.

If a shared understanding of goals is part of how we play multiplayer games, how does this influence the way we play? These three considerations can be identified in the playing of *any* multiplayer game. Danish game researcher Jonas Heide Smith has documented how players will often self-handicap in order to maintain game balance and maintain uncertainty about the outcome of a game.[6] As multiplayer games are, naturally, played in social settings, social considerations can also become part of the player's deliberations. Perhaps you do not wish to win against a fragile child. Perhaps you want to win over an annoying sibling. Perhaps you do not want to win over your boss. This means that for every choice in a game, there are three different considerations you must weigh against each other. These are as follows:

- The *goal orientation consideration*. You want to win.
- The *game experience consideration*. You want the game to be *fun* and you know that this entails making sure there is uncertainty about the outcome. You may play a little badly in a multiplayer game in order to keep the game interesting.
- The *social management consideration*. When playing with other players, you desire management of the social situation. You know the outcome of the game may make certain players sad or happy. You know the outcome

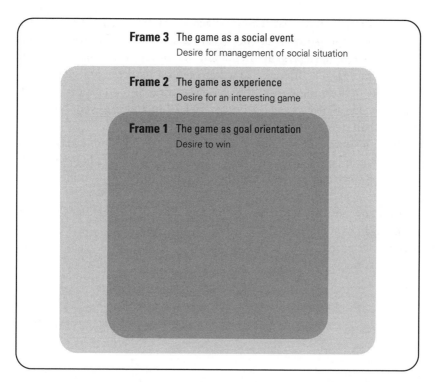

Frame 3 The game as a social event
Desire for management of social situation

Frame 2 The game as experience
Desire for an interesting game

Frame 1 The game as goal orientation
Desire to win

Figure 6.5
Three considerations for evaluating a game action

of the game may influence your social standing and the social dynamic of the group.

Figure 6.5 illustrates how these three considerations frame every game action in three competing ways. The nominal description of a game will tell you to focus on the first consideration, but the other two repeatedly come into play.

My colleague David was complaining to me: He had been avidly playing *World of Warcraft* with some friends, but by the time he'd reached level fifty-eight, far into the game, he was going through a busy spell in his life with little time to play, and his character had consequently fallen behind those of his friends. For that reason his friends refused to play with him anymore—he had become a liability. "But these are my friends," David complained. "Games are supposed to be fun!"[7] In David's experience the players disagreed about the importance of the three different frames: David emphasized social management ("these are my friends") and game experience ("games are supposed to be fun"), but his friends emphasized goal orientation, the desire to win or get further in the game. David's friends had a stereotypical hardcore attitude toward the game: performing well in the game was more important than friendship; David asserted a casual attitude: friendship and fun were more important than performance.

The uncertain interrelation between these considerations have been known to end both parties and relationships. Some players believe (at opportune times) that a friend has the obligation to help them in a game. And then again, while helping another player in a game *may* be taken to be a friendly gesture, it can also be perceived as rude and condescending. Going against someone's goals can be an affirmation that you want to be part of the social event with that person. In multiplayer games, every action has many meanings.

7   Casual Play in a Hardcore Game

Guitar Hero and *Rock Band* are often played in ways that would be hard to describe as casual, with competitions and with players putting in much effort to document their perfect performances on the "expert" difficulty setting.[1] Yet, if the player decides to play in a different way, *Guitar Hero* and *Rock Band* can also be prime examples of casual games because they are easy to pick up and play in a social situation, serving the same function as traditional social games. In this case, both games correspond well to the casual game design principles described in chapter 2: positive, pleasant fictions; easy to learn; interruptible, since a game session only lasts the length of a song; juicy, with much positive graphical feedback; and lenient punishments for failure, in that players who are *not* trying to complete or master the game can *choose* to play a different song when failing, rather than have to play the same song over and over.

In other words, these games are very different depending on what players are trying to achieve. For players who want to relax playing a song they like, the games fulfill the role of casual games where even an imperfect performance of a song on an easy level of difficulty can be a satisfying experience. However, for players who want to master the games or win a competition, the games' punishment structures match traditional hardcore design. In this case, the player must keep replaying a given song in order to perfect his or her skills. *Guitar Hero* and *Rock Band* support large time investments and reward practice with difficulty levels that scale from very easy to near impossible, providing depth as discussed in chapter 2. *Guitar Hero* and *Rock Band* are therefore not simply "casual," "hardcore," or somewhere on a scale between the two, but represent a kind of flexible design that lets players decide what type of game to play.

Figure 7.1
Sims 2 (Electronic Arts 2004)

In addition to *Guitar Hero* and *Rock Band*, in this chapter I will discuss *Sims 2*[2] (figure 7.1) and *Grand Theft Auto: San Andreas*[3] (figure 7.2). Though these four games appear very different, they are similar in that goals are less imperative than in traditional video game design. All four games are hugely popular, and even the often controversial *Grand Theft Auto* series is one of the most popular game series across age and gender.[4]

These are, on the surface, *big* games that seem to require large time investments, but they can reach a broad audience because they do not force the player into working toward an official game goal. Hardcore game design is generally inflexible toward different uses, but this inflexibility hinges in part on the convention that *players are punished if they do not work toward the game goal*. When players are free to follow personal goals, these games become flexible in terms of being played in different ways.

Welcome home, Carl. Glad to be back?

Figure 7.2
Grand Theft Auto: San Andreas (Rockstar Games North 2005)

Games without Enforced Goals

The last few decades have seen many innovations concerning goals in games: *Sims 2* has no stated goals, but is nevertheless extremely popular. *Grand Theft Auto: San Andreas* is superficially a goal-oriented game, yet it allows the player to perform a wide range of actions while ignoring the game goal. While goals provide a sense of direction and a challenge in games, they can also limit the player.

Do all games have goals? Games have changed over time, but the five-thousand-year-old Egyptian board game Senet is also recognizable as a game today.[5] Through thousands of years, games have followed what I call the *classic game model* illustrated in figure 7.3.[6] The inner circle shows the core features of games; the middle circle area contains borderline things we may argue over whether to describe as "games"; and outside the circle are things that are not considered games. Games without goals, such as *Sims 2* or *SimCity*,[7] are borderline cases because they do not have goals in a traditional sense.

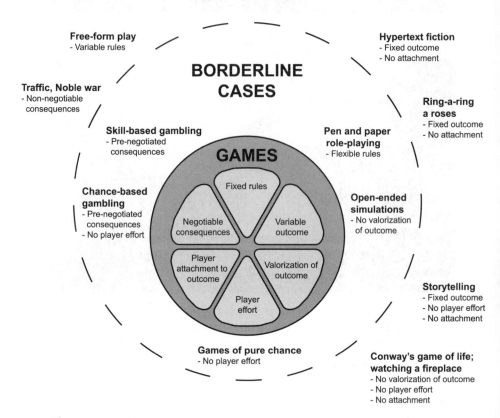

NOT GAMES

BORDERLINE
CASES

Free-form play
- Variable rules

Traffic, Noble war
- Non-negotiable
consequences

Skill-based gambling
- Pre-negotiated
consequences

Chance-based
gambling
- Pre-negotiated
consequences
- No player effort

GAMES

Fixed rules

Negotiable
consequences

Variable
outcome

Player
attachment to
outcome

Valorization of
outcome

Player
effort

Games of pure chance
- No player effort

Hypertext fiction
- Fixed outcome
- No attachment

Ring-a-ring
a roses
- Fixed outcome
- No attachment

Pen and paper
role-playing
- Flexible rules

Open-ended
simulations
- No valorization
of outcome

Storytelling
- Fixed outcome
- No player effort
- No attachment

Conway's game of life;
watching a fireplace
- No valorization of outcome
- No player effort
- No attachment

Figure 7.3
The classic game model

That a game has a goal means it is an *activity* for which some of the
possible outcomes are assigned positive values; an activity with the imperative that players *should* work toward these positive outcomes.[8] This
means that a game has to communicate its goal in some way. It also
means that soccer fields or soccer balls do not have goals, for in themselves they are not activities. The game of soccer has a goal—we use the
term *soccer* to mean a specific activity with goals. That *Sims 2* does
not have a goal means it does not tell us what we should try to achieve.
Players of *Sims 2* set their own personal goals as they wish, but "*Sims 2*"
refers to a game without goals: *Sims 2* is like a soccer ball that can be used
for a variety of both goal-oriented and goal-less activities.

Scramble: Obligatory Goals

Guitar Hero, Rock Band, Grand Theft Auto: San Andreas, and *Sims 2* are unusual in the history of video games. Compare them to the 1981 arcade game *Scramble,*9 a game that is representative of a more traditional way of constructing and communicating goals. The *Scramble* cabinet states the following:

- Object of game is to invade five SCRAMBLE defense systems to destroy THE BASE.
- Use joystick to move up, down, accelerate, and decelerate.
- Use Laser and bombs to destroy rockets, fuel tanks, mystery targets and UFOs.
- Hit fuel tanks for extra fuel for AIRCRAFT.
- Bonus AIRCRAFT at 10,000 points.[10]

The goal of *Scramble* is also indicated by way of a taunting text on the main screen: "How far can you invade our scramble system?" The explicit goal of the game is to get as *far* as possible, but there is a second implicit goal: to get as high a score as possible.

To what extent is it possible to ignore the goal of the game when playing? As a first experiment, I tried invading the *Scramble* system not as far but at as short a distance as possible. This proved very easy but not terribly satisfying: the game allowed me to simply drive the aircraft into the ground immediately three times, yielding a total score of 110 points (figure 7.4).

Since *Scramble* scrolls the screen right to left at a steady pace, the player has no option but to "invade the scramble system"—otherwise the game will end.[11] As an alternative strategy, I tried to see whether *Scramble* accommodates a pacifist playing style, when I played without attacking anything. This was impossible because the aircraft continually loses fuel, which must be replenished by attacking the fuel tanks stationed on the bottom of the screen. Failing to do so renders the aircraft uncontrollable and leads to a crash (figure 7.5).

In any prolonged playing of *Scramble,* the player is forced to work toward the goal, and to some extent to use a specific playing style. This is characteristic of how goals work in the traditional arcade game:

1. Goals are explicitly communicated.
2. There are dual goals of progressing in the game and getting a high score.

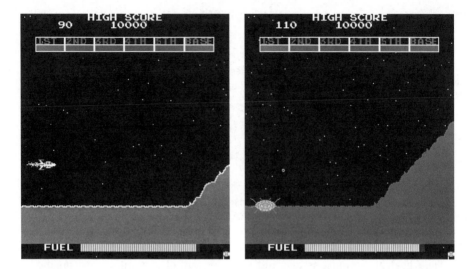

Figure 7.4
Scramble: avoiding the game goal by driving the aircraft into the ground immediately (Konami 1981)

3. The game strongly punishes players who do not try to achieve the goal.
4. The enforced goal allows the player only a narrow range of playing styles.

San Andreas: Optional Goals

The *Grand Theft Auto* games describe themselves as having goals, like the arcade game, but in actual play are very different experiences. The back cover of *Grand Theft Auto: San Andreas* states: "Now, it's the early 90s. Carl's got to go home. His mother has been murdered, his family has fallen apart and his childhood friends are all heading towards disaster. On his return to the neighborhood, a couple of corrupt cops frame him for homicide. CJ is forced on a journey that takes him across the entire state of San Andreas, to save his family and to take control of the streets."[12] The text sets up a goal not entirely unlike that of *Scramble*: the player should not "invade the scramble system," but "save" Carl's family and "take control of the streets."

Players achieve the *Grand Theft Auto: San Andreas* goal by completing the long series of missions that the game presents (figure 7.6) Again, this is quite similar to how *Scramble* works. What is not similar is that *Grand Theft Auto* does not force you into pursuing the stated goal. Right from the beginning of the game, rather than go to CJ's old neighborhood

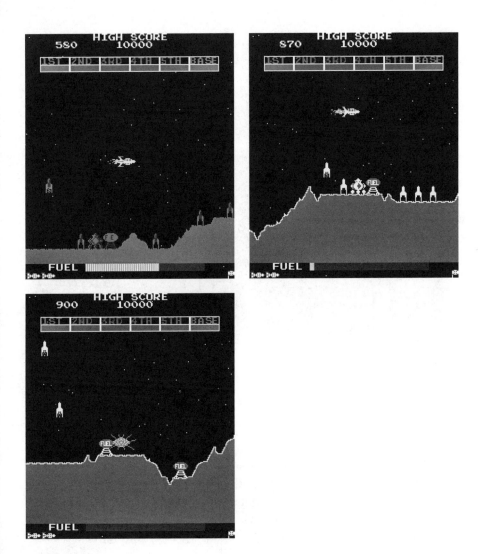

Figure 7.5
Scramble: refusing to fire any shots leads to the aircraft running out of fuel

Figure 7.6
The player is instructed to get on a bike and follow the radar to return to Carl's old neighborhood.

you can choose to bicycle around town, practice stunts, or simply explore the game world (figure 7.7).

This makes *Grand Theft Auto: San Andreas* another type of game, in which you are free to deviate from the official goal of the game and to make up personal goals such as improving cycling skills, modifying the looks of your character, or simply visiting as much of the game world as possible. *Grand Theft Auto: San Andreas* is a game with a goal, but the goal is optional.

Sims 2: Without a Goal

The back cover of *Sims 2* packaging states:

The Next Generation People Simulator
They're born. They die. What happens in between is up to you. In this sequel to the bestselling PC game of all time, you now take your Sims from cradle to grave through life's greatest moments.

Figure 7.7
Instead of undertaking missions, players can choose to explore the game world.

Create your Sims.
Push them to extremes.
Realise their fears.
Fulfill their life dreams.[13]

Sims 2 has no specific goal, but tells you that you have something akin to complete freedom—"what happens in between is up to you." In actuality, the game consists of choosing a town, creating a family or using a predefined one, making money, building family members a house, and trying to deal with their wants and desires and to make them do what you want them to. *Sims 2* does not yield full control over its characters, known as *Sims*; rather, the Sims may refuse to do what you ask them to, couples may dislike each other, and so on. *Sims 2* then pulls in several different directions: the game has no imperative and does not tell you what you *should* do, yet the Sims become miserable if you do nothing. The game is open to being played in different ways, but it also sets up a path of least resistance: purchasing more items for the Sims makes them happier, which in turn makes it easier to earn money to provide for them. Yet, a large part of the home decoration in the game is not functional but aesthetic. Sims enjoy having chairs and tables, but players can choose between a wide range of chairs and tables, many of which will be functionally identical. *Sims 2* allows the player to play for aesthetic goals. The player does not have to choose an *optimal* chair, but can choose a *beautiful* chair.

Still, you cannot do what the *Sims 2* packaging promises because much of what happens is *not* up to you: *Sims 2* is *not* a dollhouse. During my playing of the game, I instructed a character named Cornwall to eat snacks and lunch seven times in a row. I intended Cornwall to eat all the food I had instructed him to eat, yet this was not what happened. Instead, a fire broke out in the kitchen, which led to Cornwall having a nervous breakdown, and then receiving a visit from a doctor (figure 7.8).

The events in *Sims 2* are not simply up to you, but result from the interaction between the player and the game's objects, characters, and financial constraints. As such, *Sims 2* is the mirror image of *Grand Theft Auto: San Andreas*: *Sims 2* promises absolute freedom but sets up many constraints and resistances to the player's plans; *San Andreas* promises a clear goal but allows the player to ignore it entirely.

To Play without Goals

The problem with goals is that they may force us ito optimize our strategy in order to win rather than do something else that we would prefer. The last few decades have seen much experimentation with goals in games. There is a whole class of goal-less games such as *Sims 2*, *SimCity*,[14] and many role-playing games. In addition, many games *with* goals put no strong pressure on the player to pursue them. Prior to *Grand Theft Auto III*,[15] games such as *Elite*,[16] *Pirates!*,[17] and *Super Mario 64*[18] were among the more prominent games with nonenforced goals.

Games without goals or with optional goals are more *flexible*: they accommodate more playing styles and player types, in effect letting you choose what kind of game you want to play. When you are not under pressure to optimize your strategy, you have room to play for other purposes than simply winning or completing the game. Chapter 6 discussed how in multiplayer games we can choose to emphasize different considerations: we can try to win; we can aim to keep the game interesting by playing badly; we can aim to manage the social situation in which we play. If we have high social stakes in a game, we may actively try to lose if that seems advantageous. If we, for example, have bet a large sum of money on a game, we may focus on winning to the detriment of the two other considerations discussed in the chapter. It is only when we are not under strong pressure to win that we can play without pursuing the game goal.

In *Sims 2* and *Grand Theft Auto: San Andreas*, the freedom to play without following a game goal allows us to use the games for personal expression. Many players find great enjoyment in creating (and showing off) families and houses in *Sims 2*, exploring and perfecting their clever maneuvers in the *Grand Theft Auto* series.[19] One way to look at this is to think of a game as a *language*: a language contains a *lexicon* (the words) and a *syntax* (that controls the arrangement of the words).[20] *Scramble* is not an expressive game because the range of things we can do (the lexicon) is very small, and because the game forces us into playing for the goal (a very rigid syntax). *Grand Theft Auto: San Andreas* and *Sims 2* feature a wide range of things we can do (a large lexicon), while accommodating a wide range of playing styles (a flexible syntax). As such, the two games are flexible systems for expressing ourselves. In *Sims 2*, we can create houses and families that reflect who we are. In *Grand Theft Auto: San Andreas*, we can perform stunts and solve problems in ways that are unique to us. The expressivity of *Sims 2* and *Grand Theft Auto: San Andreas* also comes from the fact that these games contain elements that are already meaningful: social interaction, life and death, violence, exploration, social status, skilled performances.

The expressive power of *Guitar Hero* and *Rock Band* is quite different in that it is not visible on the screen of the game, but takes place in the space where we play. Because these games allow us to play single songs rather than try to complete the entire game, they allow us to shift our focus away from the goal of completing the entire game, to the people with whom we play, to the music itself, and to expressing ourselves by pretending to be rock stars, performing for friends and family in *player space*, outside the game screen.

The Curious Case of the Hardcore Games That Could Be Played Casually

The stereotypical casual player has a taste for positive settings, has little knowledge of game conventions, little willingness to invest time in playing, and a low tolerance for difficulty. In practice, it turns out that many

Figure 7.8
Cornwall eats snacks, prepares food

Figure 7.8
(continued)

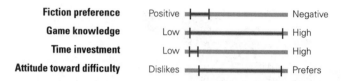

Figure 7.9
Guitar Hero and *Rock Band* as social events

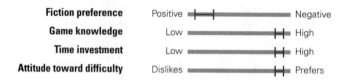

Figure 7.10
Guitar Hero and *Rock Band* as games to master

players of downloadable casual games are not like that at all, but the stereotype is useful to describe a person reluctant to play video games. Casual game design is generally flexible, affording both stereotypical casual and hardcore styles of playing, but hardcore game design typically only affords hardcore playing styles. So how do *Guitar Hero* and *Rock Band* work?

The simple answer is illustrated in figures 7.9 and 7.10: these games are different depending on the goal you set as a player. If you seek to play the game as a social event, you can play one song at a time. In this case, the game requires only marginal time investment, making it closer to casual game design principles. If, on the other hand, you play the game with the goal of mastering it, you *will* be replaying the same song over and over, and you will be memorizing the specific button presses needed to complete a song. In this case, the game is closer to traditional hardcore game design principles. The flexibility comes from the fact that the game does *not* force players to play for the goal of completing it.

That these games can be played in two ways puts pressure on the developers to cater to several different audiences at the same time. I asked Sean Baptiste of Harmonix, developer of *Rock Band*, about why they had included a "no-fail mode" in *Rock Band 2*,[21] whereby players can complete a song no matter how many mistakes they make. Though this seems like an innocuous design decision, a group of players were dissatisfied, as noted in this exchange I had with Baptiste:

JJ: *Some people feel humiliated when they fail as it stops the songs and ruins it for the entire band. Is that why you have added the new no-fail mode in* Rock Band 2?

SB: If you are somebody like me, who has played video games your entire life, for whom it has always been a hobby, then you have always been excited about showing video games to people. But the failure barrier has always been so high. The ability to demonstrate your game to people who have never played before and eliminate that barrier is a very big deal.

At the same time, we have also received some burn back from some of our more hardcore players, who were telling us, "that is not fair, they should learn how to play, life is all about failure! and so on. And we tell them "guys come on, it is a video game, it is a music video game."

JJ: *There is a certain hardcore ethic saying that "this is the way a game should be played"?*

SB: They don't want their efforts at besting the game to feel diminished by having somebody who has never played it come in and finish a song.[22]

This shows the limit of what can be achieved with game design: including a no-fail mode lowers the barrier to entry for playing a game, and makes the game more flexible, but players who believe that games should be *inflexible* and *exclusive* will not be amused. New game designs can reach a broader audience, but a given game cannot make everybody happy.

Games without enforced goals do not replace the classic goal-oriented game, but they open a wide range of new player experiences as seen in *Guitar Hero, Rock Band, Sims 2*, and *Grand Theft Auto: San Andreas*. This illustrates another strategy for reaching a broad range of players. These games are flexible in that they allow players to choose what type of engagement they want with a game. Games without enforced goals are a movement in game design that adds flexibility to large-scale video games designs.

8 Players, Developers, and the Future of Video Games

This book has tried to solve the mystery of why some people have chosen not to play video games, and why video games are now reaching new players. To those who did not play video games this was perhaps not much of a mystery in the first place—they just saw no interesting video games to play. I asked a woman in her twenties how she had come to buy a Nintendo Wii for the purpose of playing *Guitar Hero*.

Q: *Have you played other computer games or video games before?*
A: I had a Super Nintendo when I was 10. I played *Tetris* on the computer if that counts. Other than that, I didn't play anything before *Guitar Hero*.
Q: *So you had a break between when you were 10 and when you were 27?*
A: Yes.
Q: *Why do you think that is?*
A: To be honest, I am usually just not that interested in video games. But my friend got *Guitar Hero*, invited me over to play, and it was just really fun and different from other video games I have played.
Q: *Do you think* Guitar Hero *is more comparable to playing other video games or to playing a board game?*
A: To me it's not comparable to playing other video games. I view *Guitar Hero* as *Guitar Hero* and everything else as video games. So it is definitely more comparable to a board game, something you can play with friends.[1]

This player did not ignore video games because she disliked that they were *video games*, played on consoles or on computers, but because she did not enjoy the concrete video games that were available. With *Guitar Hero* she has found a game she likes, and that fits into her life. The real mystery, perhaps, is why it has taken so long for such video games to be made.

The *casual revolution* is that moment we realize that the primary barrier to playing video games was not technology, but *design*: consider how easily Solitaire card games were transformed into one of the most popular video games—it was free, usable as players already knew that rule of the game, already casual in its design, playable in short bursts; and the computer version took up even less space than the version of the game played with cards.

Let me review what *casual* means in terms of video games and players: Casual game design is based on a set of common principles that are not new in the history of games as such, but have been partially forgotten in the short history of video games. Casual players, however, are are less easily described. There is a common stereotype of casual players, but players rarely match this stereotype or the stereotype of the hardcore player. The many different ways in which we can be game players are better understood via the simple model I described earlier in this book: we have different fiction preferences; we have different levels of knowledge of video games; we are willing to commit different amounts of time; we have different preferences for difficulty in games. The stories told throughout the book and in appendix B show that players are varied and unique, but the model illustrates how players change over time, and how game designs interface with players. The diversity of game players makes it paramount that games intended for broad audiences be flexible on several levels: flexible in the assumptions about who the player is, what the player knows, how the player wants to use the game, and when. Still, a flexible game is not inherently *better* than an inflexible game. Games are, after all, rule-based structures requiring one to accept the rules that make the game activity possible. Like many forms of culture, part of the attraction of games is the prepackaged or designed nature of the experience.

It is only recently that video games have become the subject of serious study. In this nascent field there is an ongoing debate between those who focus on games and those who focus on players. This type of discussion has previously been played out in other fields. For example, in a book on film audiences, film scholar Janet Staiger tells the story of how she in her earlier work had focused on the *film*, but has now changed her mind to focus on the *viewer* and the context instead.[2] One possible future of video game studies could be a straightforward repeat of Staiger's path, first choosing games like she chose film, then choosing players and contexts like she chose viewers and contexts, then perhaps later reversing position again, replaying the movements and arguments of film theory with a delay of a few decades. Fortunately, we have another option. Thinking in terms of video games, we are free not to choose between games and players in the first place: since there is no scale by which we can measure the role of the player (and context) against the role of the game, any claim of one being *more* important is highly dubious. In fact, it is clear that while different players use and understand a given game differently, players also have preferences for different games, so neither players nor games can be ignored. Casual games are a clear indication that game design matters, but it matters in part by being flexible toward different uses. It is therefore meaningless to try to emphasize game over player or player over game. The hope is that I have managed to avoid that trap here, managed to give more equal weight to players, games, and developers too, and to demonstrate how this yields an understanding of casual games and casual players that cannot be achieved any other way.

Why Casual Games Now?

Why casual games *now?* While I do not think that casual games were inevitable, we are also experiencing something of a perfect storm of combining factors that is aiding the rise of casual games. One factor is the simple issue of changing demographics: the first generation of people who played video games as children are now well into their forties, have less time on their hands than they used to, but are looking for video game experiences that work for them today. Game developer Jacques Exertier predicts a second boom in game development for senior citizens in a few years' time.[3]

The less obvious second factor is the widespread presence of personal computers in the industrialized world. A woman in her seventies told me how she came to play downloadable casual games: "In 2000 I acquired a computer because one of my sons moved to Italy with his wife and child. I wanted to email them. Some friends showed me a computer-based Mahjong game, which I found amusing since we had played it on a beautiful set in my childhood home. Shortly after, I found some games in a CD-ROM magazine from the library where I worked. This included Mahjong, *Yahtzee* and various puzzles. That was how it began. The next game was *QBeez*, which I got completely hooked on, and so on."[4] This player probably would not be interested in buying a console for playing games, but she had a computer, and through that computer she had access to video games she enjoyed.

The third and final factor is the pure economics of video game development. In chapter 1 I discussed Microsoft and Sony's promises for the current generation of consoles, boasting about improved graphics as their most important new feature. During the last thirty years, game development budgets have been doubling for every new console generation, and roughly doubling the amount of games needed to be sold in order to make back the money spent to develop a video game.[5] The current top retail video games have typical budgets of around $15 million and need to sell one million copies to make that money back. If these trends were to continue, a video game released in the year 2064 would have to sell a billion copies in order to be profitable. This is an unlikely scenario, so what will happen instead?

The basic facts are: the development cost of a game does not translate directly into value for the player, and growing video game budgets appear to be reaching a point of diminishing returns. The economics of video game development are already quite uneconomical, so to speak, as players often do *not* finish the games they've purchased. *Half-Life 2: Episode One*[6] is a short game by modern standards, taking only around five hours to complete. Nevertheless, only 40 to 50 percent of players actually complete the game.[7]

Because of their smaller scope, casual games are generally cheaper to develop than the larger hardcore games that have driven the video game industry for so long. The rise of casual games shifts the perspective from technical graphical fidelity to more mundane questions such as: how *does* a game fit into the life of a player, and how much meaning can the game acquire from the context in which it is played? Casual games are an alternative answer to the old question of how to make games that players feel are worth their time and money.

Games Everywhere

I have focused on mimetic interface games and downloadable casual games in part because they are tied to identifiable commercial distribution channels with identifiable actors. The casual revolution also contains a less easily traceable trend of small browser-based games such as the widely popular *Desktop Tower Defense*[8] (figure 8.1). If downloadable casual

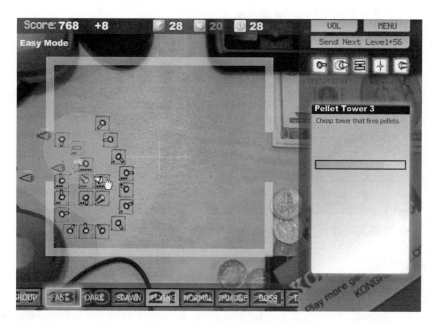

Figure 8.1
Desktop Tower Defense (Preece 2007)

games are usable in part because they only take a few minutes to download and install on equipment the player already owns, browser-based games take it a step further by being playable within a few seconds of simply going to a web page.[9]

Social networking sites such as Facebook provide another opportunity for playing video games, as well as for the social embedding of games and creation of simple games that are interesting *because* they are played in a social context. Like the example of *Parcheesi*, the Facebook game shown in figure 8.2, called *Parking Wars*,[10] is nominally a simple game in which players gain points (money) by parking their cars on the "streets" of other players in the game. From a purely strategic standpoint, the game does not offer much at first, but what makes it interesting is that the streets available primarily are those of Facebook friends, giving players the feeling of visiting friends, and of being not so nice to them by fining them when they park on *your* street. *Parking Wars* does have a final twist on the description of casual games as flexible, as this is an *inflexible* casual game: in the first few months of playing the game, the

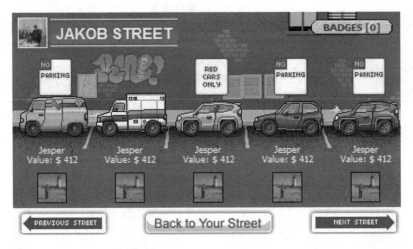

Figure 8.2
Parking Wars: exploiting that another player is on holiday (area/code 2008)

near-optimal strategy is to play the game for a few minutes twice a day; the game does not support a more intensive time commitment.

Chapter 4 showed how the mechanic of matching tile games has become so well known that it is easily added to other, larger games. Facebook games like *Parking Wars* demonstrate that the idea of playing a low-intensity video game is now so well understood by developers and players that games of this kind can easily be introduced as minor components of larger activities, such as social network sites.

Two Tales of Casual Games

There is a happy story to tell: with casual games, video game developers have stopped making games just for themselves and are reaching underserved audiences, and a much larger part of the population is experiencing the joy of video games firsthand. The video game medium is blossoming and finally reaching the audience it deserves.

It may seem hard to have issues with this, but some people do: "If gaming is to be accepted as an art form, should the complexities really be ironed out so Grandma has more of a chance to come first in Mario Kart?" And: "What if the Church asked Michelangelo to paint the Sistine Chapel as a cartoon so the masses would find it more accessible? It's blasphemy."[11]

This shows that we can tell the story another way: a group of people cared deeply about the art form of video games only to see it kidnapped by commercial interests that want to simplify and dilute video games in order to sell them to a broad public that doesn't know much about video games. In this case, it is a story about art cinema vs. Hollywood, experimental literature vs. pulp fiction, experimental bands vs. the music industry. Some developers and players are unhappy with the way things are going. Some developers feel that it is becoming hard to create interesting content for casual games.[12]

The worries of the traditional video game players and developers are thus based on very real changes in video games, the video game industry, and the video game audience. There are conflicting perspectives here: the person who was not playing video games suddenly has access to games that he or she finds interesting; the developer of traditional hardcore games is feeling economic pressure toward making games that reach a larger market; the traditional hardcore player is worrying that the games he or she enjoys will stop being made. For some players, there *is* a genuine sense of loss, watching games becoming mainstream and accessible.

Part of the happy story was also that, finally, video game developers were no longer making games "for themselves."[13] In a broader perspective, making cultural forms for an audience other than oneself often is derogatorily described as "selling out." In fact, creators of even the most popular media products often go out of their way to assure us that they certainly *never* thought of making anything popular, but only tried to make something that they themselves would enjoy. J. K. Rowling, author of the *Harry Potter* series, explains that she did not write with the ambition of the books becoming bestsellers, but that she wrote *for herself*: "When I write the books, I really do write them for me. Very often I get asked, 'Who do you have in mind when you write? Is it your daughter or is it the children you've met?' No. It's for me. Just for me. I'm very selfish: I just write for me."[14]

The cultural status of video games remains problematic: it is as if video games are always considered more dangerous than other forms of culture, and that developers are therefore under pressure to err on the side of acceptability. It could be called video game exceptionalism—the idea that video games are just *so* special that normal assumptions about art, expression, and culture in general do not apply.

With casual games, we can hope that as video games become ever more *normal*, it will be widely understood that they are a legitimate form of cultural expression, a kind of cultural expression that comes in many shapes and sizes.

A Casual Revolution

My goal in this book has been to talk about a specific time in video game history. Sometime in the future, perhaps much of this discussion of a changing video game audience will seem quaint, like the talk of living "at...railway speed" in chapter 3. This book documents that cultural moment when video games became normal; when it is no longer exceptional to play games using computers and televisions; when the emphasis has shifted from *"video* games," to "video *games"*: as I write, video games are becoming but one type of games among all the others in our culture. Paradoxically, this normalization of video games helps them fulfill their potential by making it absurd to ask whether we are for or against video games, and thereby making it easier to appreciate the nuances of specific games as well as future experiments and experiences.

It has become clear why, before the rise of casual games, some people did not feel the pull, the attraction of video games. In this way, casual games have made the conventions of traditional hardcore games visible. We can now see how mistaken assumptions have held video games back by focusing on emotionally negative situations, by presupposing encyclopedic video game knowledge, by requiring intensive time investments, by punishing players needlessly. We can also see that the negative prejudice equating video games with awkward young males has kept potential players away from games they might have enjoyed. Over the last few years, we have learned to create video games that reach that broad audience. Not that everybody is playing video games yet, but there is nothing that prevents this from happening. Video games are fast becoming games for *everyone.*

Appendix A: Player Survey

To develop a clearer picture of the players of downloadable casual games, I conducted a survey in the fall of 2008. I recruited the players in collaboration with Gamezebo, a website that publishes reviews of and news about downloadable casual games. On September 2, 2008, Gamezebo published the following news item:

Jesper Juul, a researcher at the Singapore-MIT GAMBIT lab in Cambridge, Massachusetts, is writing a research study and book on casual games and wants to interview Gamezebo users to get your opinions.

    Mr. Juul's research will seek to answer the questions of who plays casual games and why they are so popular lately. The online survey is voluntary, anonymous, and should only take 10–15 minutes to complete.

The survey yielded statistical results discussed in this appendix and a number of life stories discussed in appendix B. The statistical results were mostly consistent with the data from other surveys, but there is an important caveat concerning what they show: media scholar Jason Mittell has discussed how in audience research it can be ambiguous who exactly qualifies as being "the audience" for a given TV show, for example.[1] Casual players present the same type of problem: who are the casual players? Are they those players who regularly visit casual game websites? Those who regularly *buy* casual games? Everyone who has ever played a casual game? Players who play at least once a week? This survey is similar to previous ones in that it is likely to be a survey of the most dedicated players of downloadable casual games: recruiting players via a casual game website such as Gamezebo is likely to generate responses primarily from dedicated players, since less-dedicated players rarely visit these sites.

Since answering the questionnaire was voluntary, that factor too is likely to recruit the most dedicated players.[2]

The survey included several questions about casual games. Here, the term *casual games* is meant to signify downloadable casual games only, excluding mimetic interface games. In order to minimize the impact of different understandings of the term, players were specifically asked for their opinions on "casual games (games such as the ones on the Gamezebo website)."

One hundred eighty-two users fully completed the survey. These are the key results:

- The average age of the respondents was 41, with the youngest users in their teens and the oldest in their seventies.
- 93 percent of the respondents were female. This may have to do with the composition of the Gamezebo audience, but that is speculation.
- 98 percent play casual games at home, 2 percent played them at work.
- 35 percent play several times a day for at least one hour.
- 14 percent play several times a day for more than three hours. This translates to more than thirty hours a week, forty-two hours if we include weekends. This is much higher time usage than the casual player stereotype suggests, but the results are similar to those of previous surveys.
- Players were keenly aware of genre categorizations, as illustrated in figure 3.2.
- To the question, "How important is it for you to be able to pause or shut down a game if you are interrupted?" 81 percent answered "Very important," 18 percent answered "Somewhat important," and only 2 percent answered "Unimportant." Interruptibility is important to most of the players.
- In terms of difficulty, players were asked, "Which is worse: a game that is too hard or a game that is too easy?" While 48 percent reported "Both are equally bad," 17 percent reported that a game that was too hard was worse, whereas 29 percent reported that a game that was too easy was worse. Again, the stereotype of casual players predicts that they dislike difficult games, but the surveyed players skewed toward hard games.[3]
- Casual games are often claimed to be different than more traditional computer/video games. The survey included questions to explore how players conceive casual games in that respect. When asked "In your opinion, are casual games (games like the ones on Gamezebo) a kind of computer game?" 95 percent of surveyed players answered yes.

• When asked "In your opinion, are casual games (games like the ones on Gamezebo) a kind of video game?" 56 percent of surveyed players answered yes. Players were also given to opportunity to provide text responses to this question. The answers demonstrated how some players distinguished between casual games and "video games" by way of both platforms and content, such as, "To me, a video game is one that is played on another type of system, for example PS or Wii. I also tend to associate 'video game' with more arcade style or action games, though I know this is not the case." Some responses asserted that computer and video games could not be told apart: "Aren't computer games the same thing as video games?"

I was also interested in whether players perceived any barriers to playing "traditional computer/video games" (and if players perceived any difference between casual games and the more traditional computer/video games). The question was, "Is there anything that makes you choose a casual game over traditional computer/video games?" Players could select more than one option, as follows:

Computer/video games are too expensive	44.9%
I don't have the time to play computer/video games	23.7%
Computer/video games were not made for me	21.8%
Computer/video games are too hard to pause if I am interrupted	21.8%
I don't like the stories of computer/video games	19.2%
Computer/video games are too frustrating	18.6%
I think computer/video games too hard to figure out	15.4%
Computer/video games force me to play the same thing over and over	14.1%
Computer and video games don't have nice effects	6.4%
I don't like the graphics of computer/video games	5.1%

In order to learn about the process of selecting a game to play, players were asked "What makes you want to play a casual game?" They could select more than one option:

If it is a game type I know and like	86.3%
If it is a new variation on a game I know	78.1%
If it is challenging	69.9%

If I like the story	57.4%
If I like the graphics	57.4%
If it is an entirely new type of game	57.4%
If it is relaxing	56.3%
If I can play it in small breaks	50.8%
If there is good variation between levels	49.7%
If I like the sound	13.7%
If it is similar to a non-computer game that I already know	15.3%

These results tell us something about the most dedicated players of downloadable casual games. It is predictable that such studies of casual players will yield results about their "surprising" dedication: the stereotypical casual player indeed has a "casual" relationship to games, but the selection bias inherent in having players voluntarily fill out surveys on casual game websites all but guarantees that an actual stereotypical casual player would be unlikely to participate in the survey. What we *do* know is that the most dedicated players of downloadable casual games are, indeed, extremely dedicated.

Appendix B: Player Stories

I make a habit of asking people I meet about their video–game-playing habits (or lack thereof). When I began working on this book, I noticed the following three types of player stories, mentioned in chapter 1:

1. The player who played video games as a child and teenager, then stopped, and is now returning to playing video games.
2. The player who was a dedicated "hardcore" player and now has less time to play and therefore has begun to play casual games.
3. The player who has never played video games before and is only now discovering casual games, and video games.

While writing this book, I collected more player stories: the survey of players of downloadable casual games discussed in appendix A included an open-ended question of about how players' game-playing habits had changed over time. I also conducted e-mail and phone interviews with players of both downloadable casual games and mimetic interface games. During this research I became aware of a fourth type of player story:

4. The player who retires, or whose children move away from home, or who becomes disabled and therefore has more time for playing games.

The following stories tell how players came to play casual games, divided into the four story types. Finally, I have included a few stories that did not fit into any category. Player stories from the survey are reproduced at full length, player stories from interviews are presented as excerpts.

Returning Players

Type 1: These are the stories of players who played video games in the past, especially as children, took a break from video games, and now have returned to video games via casual games.

Interview with a 27-year-old woman who plays only *Guitar Hero* and has bought a Nintendo Wii for the purpose of playing that game.

Q: *Do you play computer games or video games other than* Guitar Hero?

A: Not really. I played *Wii Sports* that came with the Wii.

Q: *Have you played other computer games or video games before?*

A: I had a Super Nintendo when I was 10. I played *Tetris* on the computer if that counts. Other than that, I didn't play anything before *Guitar Hero*.

Q: *So you had a break between when you were 10 and when you were 27?*

A: Yes.

Q: *Why do you think that is?*

A: To be honest, I am usually just not that interested in video games. But my friend got *Guitar Hero*, invited me over to play, and it was just really fun and different from other video games I have played.

Q: *Do you think* Guitar Hero *is more comparable to playing other video games or to playing a board game?*

A: To me it's not comparable to playing other video games. I view *Guitar Hero* as *Guitar Hero* and everything else as video games. So it is definitely more comparable to a board game, something you can play with friends.

Q: *What about the difficulty in* Guitar Hero? *Are you the kind of person that wants to beat* Guitar Hero *at expert level, or how is that?*

A: I do care. I think the difficulty is pretty good. I am trying to transition to hard right now, and I find it difficult. But it's nice because there is always a challenge, so you don't get bored with it.

Q: *Do you ever play card games or board games or competitive sports?*

A: Not really.

Survey response from a 38-year-old female player.

Q: *Have your game-playing habits changed over the years?*

A: The only video games I played before were *Tetris* and *Sonic the Hedgehog*. Even after I got my first computer, I didn't play many computer games. It was only after finding "Casual Games" that I really started playing computer games consistently.

Survey response from a 25-year-old female player.

Q: *Have your game-playing habits changed over the years?*

A: I used to play console games when I was younger, but now I only play casual computer games.

Survey response from a 47-year-old female player.

Q: Have your game-playing habits changed over the years?

A: I used to play console/computer games, but I need to purchase/update systems before I can continue to play those games. I found casual games because I was looking for something more affordable in the meantime. Now, I thoroughly enjoy them as an entertainment source in their own right.

E-mail interview with a male game player in his twenties who currently does not own a game console.

Q: Have your game-playing habits changed over the years?

A: Previously my younger brother and I shared a Super Nintendo, N64, and GameCube, but when I moved out he kept the systems. I have insanely fond memories of playing Super Mario World on the SNES the year it came out. My brother and I got the system (packaged with the game) for Christmas and I think we spent most of our Christmas break from school waking up, getting heaping bowls of Rice Krispie Treats cereal and sitting on the floor in the living room in our pajamas playing the game. I think due to these fond memories it's my favorite console, but I really loved the N64's controller. . . . I largely stopped video games when I reached High School because I simply ran out of time: several AP classes, a lead in the school musical, coeditor of the yearbook, officer in the drama club, an art student prepping a portfolio and a final film, on top of the inevitable college search process. It just left no spare time for myself. And at the time, I didn't have any close friends who played video games, so if I was hanging out with friends, we weren't gaming. When I went away to college I didn't have a TV, so I didn't bring any consoles either. *Rollercoaster Tycoon* on my PC, however, filled in [the] gaming niche. I really WANT to buy a Wii, currently, as *Rock Band* and games like *WarioWare* and *Wii Sports* have been uber-fun to play and I can definitely see them fitting nicely into my current life. But sadly, still, I need to buy a TV first.

E-mail interview with a female player of downloadable casual games in her mid-fifties.

Q: How did you first find out about casual games? (By casual games I mean "games such as the ones Gamezebo writes about.")

A: The first casual game I really got into was *AstroPop*, which I first played on the *PopCap* website, and then downloaded. For a long time, that was the only game I played. Early on, I also downloaded (and still play) a *Puzzle Express* game, in which you fit different-sized blocks into railroad cars. Then I stumbled across *Diner Dash*. This was the portal through which I entered the gaming world.

Q: If you were to describe casual games to somebody who doesn't play them, how would you describe them?

A: [As g]ames played on the computer in bite-sized pieces, which don't require learning and retaining complex rule sets or a great deal of data. Often they have a simple story line, which enhances, but is not central to, the game action.

Q: How often do you play casual games? For how long do you play at a time?

A: I play casual games almost every day unless I'm traveling or involved in a special project. When I'm at home I usually play them for two or more hours a day. I will often put in an hour and a half in the morning, and then a shorter stint in the afternoon.

Q: What other games do you play? Computer-based games? Console games? Cell phone games? Card games? Board games?

A: I do occasionally play cell phone games such as *Tetris*. I play cards when traveling and board games at family events. I do not play console games currently, but long ago I was very good at *Tetris* on the early Game-Boy system.

Q: If I understand correctly, you had a break from computer/video games for a while. Can you tell me the brief story of how your game-playing habits have changed over the years?

A: My first encounter with electronic gaming came when I came across an abandoned *Mario Bottle Factory* game. This was a hand-held electronic game "dedicated" to one game only; in other words, not a game system. I became hooked on helping Mario fill his bottle orders! After I had mastered that game, I heard of a separate, "dedicated" hand-held sequel, *Mario's Cement Factory*. I played that for months and never conquered it. A couple of years later, my sons, aged then about 14 and 11, got Game-Boys—the first model, bulky and monochromatic. My favorite game on the GameBoy was *Tetris*, and I played it mostly when we were on family trips. I bought two sequels to *Tetris*, but only liked one of them, whose name I can't now remember. I beat *Tetris*'s highest level, and also, I think, beat the sequel I liked. But after that I just drifted away from electronic games. I have always worked on computers, doing graphic design and word processing at home on a pro bono basis for nonprofit organizations. And of course I had an Internet provider. One of my kids discovered *Fruit Smash* on the MiniClip website. I got started playing *Fruit Smash* and then stumbled across the PopCap website where I became addicted to the online version of *AstroPop*. Once I had beaten all the online levels, I took the plunge; I bought *AstroPop*, my first downloadable game! Soon after, I downloaded *Puzzle Express*, but then stayed in a holding pattern till *Diner Dash* came along. I bought that and played it every day till I beat it, then went back and played it again. I had found my favorite game genre: time management with a female protagonist. I liked seeing Flo get ahead and also very much enjoyed fixing up the run-down restaurants as the game progressed. My favorite *Diner Dash* of the first three was the second, in which you could choose your own renovations. I especially liked that getting an "Expert" rating on a level actually meant something in that game; it allowed access to decorating resources unavailable for those who had merely passed the level. My next discovery was BigFish games. I loved the idea of a new game every day, and I soon became a member. I check it every day I'm not traveling, try about four games a week and buy one every week or so. Several months ago I discovered Gamezebo and check that every day as well. I mostly use it to discover what new games are out, read their reviews, and follow the links to the game downloads.

Q: *Have you tried games like the Nintendo Wii games*, Guitar Hero, or Rock Band?

A: I have tried both the Nintendo Wii system and *Guitar Hero*, which my sons love.

Q: *What do you think about these kinds of games? Do you definitely prefer downloadable casual games, or do you also like to play the Wii games or* Guitar Hero?

A: I vastly prefer casual games. I am familiar with the keyboard/mouse computer interface and don't want to be bothered learning another controller.

Q: *What is more important in a casual game: story, graphics, sound, the right difficulty, or the right amount of variation?*

A: The right difficulty. And for me, the right type of action. I prefer time management games and hidden object games. Also, because of religious reasons, I can't buy games that involve the occult at all. I wish developers, especially hidden object developers, would stop making games with occult-based story lines!

Q: Would these games be more enjoyable if they didn't have levels that you get stuck on, or is getting stuck (and then getting further) part of what makes these games enjoyable?
A: If a game has a level, like at the end of a stage, that you have to play and replay to beat, it's an acceptable and even enjoyable challenge. But if you move through all the levels smoothly and at the very end encounter one that has increased exponentially in difficulty to the point of being impossible, it is a total turn-off. I think that the key phrase in your question is "and then getting further." Having a level so difficult you really can't get any further alienates me COMPLETELY!

Q: Can you assign a number to how many times it is still enjoyable to retry a level before it becomes "too much," or does it depend on the game?
A: I would say that if your skills have been honed by a steadily increasing learning curve, then ten to twenty would be the maximum replay number.

Ex-Hardcore Players

Type 2: These are the stories of players who used to intensely play video games and now have switched to more casual video games.

Survey response from a 40-year-old female player.
Q: Have your game-playing habits changed over the years?
A: I used to only play RPGs like *Guild Wars* but you can start and stop casual games easier during the day.

Survey response from a 42-year-old female player.
Q: Have your game-playing habits changed over the years?
A: Started with text-only adventure games, moved toward RPG video-games & simulations, most recently I stick with time management-type casual games.

Survey response from a 29-year-old female player.
Q: Have your game-playing habits changed over the years?
A: I no longer play shoot 'em ups or beat 'em ups or two-player games with my sister on the Amiga. I've less patience with poor games and am less inclined to persevere. My shelf's full of games I've bought then never even bothered to play, or those I've only played for an hour then given up. At least with casual games the free trial makes that less likely. I've always

played casual games, even before they were called that though (*Tetris*, pinball, card games, *Nuclear War, Rockstar Ate My Hamster*, they were all casual) and I've always played traditional games too.

Survey response from a 38-year-old male player.
Q: *Have your game-playing habits changed over the years?*
A: As I grew up and had more obligations my time and patience became limited towards investing in epic games. Though I still love the idea of playing epics like *Civilization* or *Warlords* or *SimCity*, the time required is just more than I can provide. Every so often I try to get a game going only to be pulled off it by various obligations and [I find] it difficult to return.

Survey response from a 30-year-old female player.
Q: *Have your game-playing habits changed over the years?*
A: Having a baby really changed my game playing habits. When she needs my attention the game must stop. This is why *World of Warcraft* has been hard to play as of late.

Survey response from a 43-year-old female player.
Q: *Have your game-playing habits changed over the years?*
A: I've been an active computer gamer since 1989. I've always loved the adventure games. But as I've grown older, got married, had kids, I find it hard to concentrate too long and get too involved in an adventure game, since the time that I spend on the computer is so inconsistent. A casual game is now perfect for me...it helps me to relax and "stimulate the gray matter." I love them!

Players Discovering Casual Games

Type 3: These are stories of players who have discovered video games through casual games.

Phone interview with the father in a Wii-playing family, the parents in their early thirties with two twin girls aged three and a half.
Q: *You compared the Wii to* Parcheesi?
A: We don't play *Parcheesi* [*Sorry!*/*Ludo*] with the kids, because it is too complicated for them—they are only three and a half years old. With the Wii, on the other hand, the way that you do something and see a reaction

on the screen, the way you tilt the controller and see something on the screen—that is something different. You cannot give them PlayStation controllers; those are a little too advanced with too many buttons. With the Wii, we can see on the kids that it just works for them, they can use that immediately. We play the Wii with friends, at social events. We have also played it with the in-laws who are both around sixty. They play it eagerly, and they ask if we shouldn't play the game one more time.

Q: *Do you personally play other computer or video games?*

A: Ah yes. I have started playing *Call of Duty*, and I used to play *Counter-Strike* a lot. I am into first-person shooters, we have a clan, and so on. But nothing related to the Wii.

Q: *You haven't tried converting your wife or family to computer/video games?*

A: Not to traditional computer games. I know they don't like those, so it hasn't come up. We play the *Settlers* board game with the in-laws. The computer is not so good for something like that where it becomes strategic and you play for several hours. When I was a child, we played *Parcheesi* and chess, or perhaps *Pong*. That could be played with the family.

Q: *What Wii games do you play?*

A: Mostly *Wii Sports* and *Wii Fit*. We have bought some others, but we don't play them. We just held a summer barbecue with eighteen guests. Everybody was playing the hula hoop on *Wii Fit*. We bring out the Wii at social gatherings and when friends come over.

Phone interview with a player of downloadable casual games in her fifties.

Q: *Have you played board games or card games?*

A: Lots. Checkers, Nine Men's Morris, and lots of card games.

Q: *And Solitaire games?*

A: Yes. Playing casual games actually feels similar to playing Solitaire. You are totally relaxed, you cannot concentrate on anything else, but at the same time you can be thinking about other things in the back of your mind. I often play when I face a difficult problem. In my company I face various tasks that are hard to get started with. I already have the knowledge I need, so I play a game rather than go read a lot of books. Then the solutions come. It is like the game brings out a lot of tacit knowledge, as if the problem solving in the game maintains that skill, and that is a skill I need.

Q: *How were you introduced to casual games?*

A: My 75-year-old friend introduced me to *Zuma* and *Collapse*, the predecessor to *Zuma*. It was after I had handed in my PhD thesis, so my brain was completely offline. Then she invited me over for dinner and told me she had something interesting to show me. She also had a computer Mahjong game that was very beautiful and exciting, I really liked that. Later I have begun to buy them myself, because they are not that expensive.

Q: *How do you feel about difficult games? Is it a problem to be stuck on a level?*

A: Level twelve of *Zuma* is really fast. I think I gave up after fifty attempts. *Zuma* has a game mode called Gauntlet where you can practice different levels, so I switched to that and practiced becoming faster. That helped, but I was still too slow. It was important for me to finish the game—I believe that is important in life, to finish things, no matter what. I like competing with myself, to see development and progress. "No matter what," is really the point for me. I googled for solutions and found a site with a cheat code to make *Zuma* slower. It worked!!! For me, that was even more satisfying that beating the game on its own terms: to modify the game to fit my own limitations and capacities.

Survey response from a 52-year-old female player.

Q: *Have your game-playing habits changed over the years?*

A: Until I discovered casual games on the computer I used to spend a lot of time with traditional crossword puzzle books and other puzzle-based paper-based activities.

Survey response from a 49-year-old female player.

Q: *Have your game-playing habits changed over the years?*

A: I play more now that they have made games to suit women. Not the fighting, killing, kicking...etc. games.

Survey response from a 29-year-old female player.

Q: *Have your game-playing habits changed over the years?*

A: I only discovered 'casual games' about a year and a half ago. Of all things, my Mom had bought *Insaniquarium* and a puzzle type one (I want to say *Penguin Puzzle*, but I don't remember the name for sure) for my son for Christmas. Regardless, after the entire family got hooked on *Insaniquarium*, I ended up checking out the website of the company that put it out, and things went from there. Before 'casual games' entered

the house, I'd gotten to the point where I mostly played MMORPGs—
EverQuest, at the time, though I usually ended up giving new ones a try
as they came out. That, and *Sims 2*. But like I said above, 'normal' computer games don't come out all that often. At least, not ones I was interested in. With the whole new world of casual games that can be
downloaded and tried in just a few minutes, it let me have a much wider
variety of games to play, so that I now have something to play no matter
what my mood is and what I want to do.

Players with Life Changes

Type 4:   These are the stories of people who did not play video games before, but who due to changed life circumstances find themselves with
more time to play video games, especially casual games.

Survey response from a 68-year-old female player.
Q: Have your game-playing habits changed over the years?
A: As I have aged and my children are gone and I am only working
part-time, I play more than I used to. I believe it will also help to keep
my mind active and thinking.

Survey response from a 56-year-old female player.
Q: Have your game-playing habits changed over the years?
A: I play much more than I used to as I have become disabled and have
a lot to hours to fill. I buy very few games. I belong to Shockwave where I
have unlimited play while connected to the Internet. I purchase games
with repeatability to have on the laptop when traveling or unable to access
the net.

Survey response from a 48-year-old female player.
Q: Have your game-playing habits changed over the years?
A: Kids are all grown up and have left home and I have more time to
play casual games. I find them relaxing and a way to de-stress.

Survey response from a 40-year-old female player.
Q: Have your game-playing habits changed over the years?
A: Being disabled two years ago I spend most of my time with playing
casual games.

Survey response from a 61-year-old female player.

Q: *Have your game-playing habits changed over the years?*

A: Since I have retired I play more of video and casual games.

Survey response from a 21-year-old female player.

Q: *Have your game-playing habits changed over the years?*

A: I've started playing more in the last two years while I've been a stay-at-home mom.

Other Stories

Finally, a number of player stories do not fit into any other category.

Survey response from a 51-year-old female player.

Q: *Have your game-playing habits changed over the years?*

A: Having had viral encephalitis in 1998, I lost short-term memory. I needed a way to challenge my memory and work on visual, tactile, and hand-eye coordination. I was a medical student before the brain injury. I understood the "use it or lose it" theory of the brain. These games seemed the perfect therapy. I feel they have allowed me to keep what is left of my brain functions sharp.

E-mail interview with a female player of downloadable casual games in her mid-fifties.

Q: *How did you first find out about casual games? (By casual games I mean "games such as the ones Gamezebo writes about.")*

A: Through the different "Play free for 30 minutes" places on the net. I've always liked non-arcade computer games. I think I put Sierra's Roberta Williams' kids through college!

Q: *Since you mention Roberta Williams, it sounds like you have been playing adventure games?*

A: Yes, I started in the early 80s with almost everything in the early Sierra Online line. They were truly unique at that time, as it seems to me that Sierra was the first software house that realized females were playing games too, and so Sierra populated their games with female characters. Very cool at the time. At that time, it seemed very few women were involved with computers. The only reason I got into them was as a tool to write a seminar on cognitive theory with a friend of mine. Who knew?

Q: Can you tell me the brief story of how your game-playing habits have changed over the years?

A: Wow....A lot more choices. Initially, I just played everything that was available—arcade, various point-and-click, "Myst"-type. About fifteen years ago, a car accident wrecked up my hands a bit, and so now I stick with so-called casual games, and adventure, puzzles, etc. While I was able to go back to work as a graphic artist, the lightning-quick reflexes were kind of shot.

Q: If you were to describe casual games to somebody who doesn't play them, how would you describe them?

A: They're games (obviously) that make you look at things in different ways....They involve problem solving, looking at things out of context, and often using both the right and left sides of your brain.

Q: Before playing, did you have prejudices against playing games on computers (including casual games)? Did you experience other people having such prejudices?

A: No; I used to work in the computer industry as a tech writer, and, by the time online gaming had started, I had most of my home on my computer—checkbook, grocery shopping, cooking, etc. It was a pretty natural progression for me. After I left the computer industry, I would still do beta-testing on apps that interested me. ("Will beta for software!") What I did experience was a lack of games geared towards women and girls; that seems to be changing now, which is nice.

Q: What other games do you play? Computer-based games? Console games? Cell phone games? Card games? Board games?

A: "Thinking"-type board games, trivia, bridge, chess.

Q: Have you tried games like the Nintendo Wii games, Guitar Hero, or Rock Band?

A: No, we don't have a console. I also find that type of game gets boring after a while. Although I sometimes play *Dance Dance Revolution* with my teenaged nephew.

Q: How often do you play casual games? For how long do you play at a time?

A: Daily. How long depends on why I'm playing. If it's to take a break between doing stuff around the house, about fifteen–twenty minutes a pop. If it's to distract myself or relieve stress, maybe an hour. If it's to zone out, it could be several hours.

Q: What is more important in a casual game: story, graphics, sound, the right difficulty, or the right amount of variation?

A: Honestly, for me, with certain types of games (hidden object, for instance) story line isn't very important, unless it's a good one! Ditto sound; after the first few minutes, I generally turn them way down. Probably variation is most important, followed by the degree of difficulty. The game's integrity is very important, and its consistency.

Q: *Could you elaborate what you mean by "integrity"?*

A: This is a big complaint of mine. In hidden object games, for instance, there will be a murky shiny, round object. One round it's deemed to be a metal coin. The next, it's miraculously transformed into a locket. Or a medal. Or there are four buckets in a scene, but the developers want you to click on a specific one...a blue bucket, say, and ignore all the other buckets you try, even though the "Find" list didn't specify "blue." Or a character suddenly acts totally contrary to how they've been acting all along, with no explanation given. By integrity, I don't mean anything big, like, politically. I just mean integrity within the confines of that particular game. Puns, both visual and word, are fine, and tricky wording is fun, but those examples above go over the line.

E-mail interview with a self-described female casual gamer aged 29.

Q: *What games do you play?*

A: I'm a casual console gamer—Wii and Xbox (and Xbox 360)—and I mostly play role-playing games. I love open-ended RPGs like the *Elder Scrolls* series (or rather, those two for the Xbox, *Morrowind* and *Oblivion*), as well as the more plot-based games put out by BioWare, like *Star Wars: KOTOR, Jade Empire, and Mass Effect*. *Fable* I enjoyed as well, and I'm looking forward to playing *Fallout 3* when it comes out (and I have the time). But I'm not a *Final Fantasy* person. I'll also occasionally play FPS games, like *Perfect Dark Zero* and *Halo*, though usually the learning-curve is too steep for me, and I like play *Wii Sports* with my husband, as well as the odd stealth game (*Thief: Deadly Shadows* is one I've played through more than once, and I've been wanting to play *Assassin's Creed*).

Q: *When you describe yourself as a "casual console gamer," what do you mean by that? What makes you "casual," rather than "hardcore"?*

A: When I call myself a "casual gamer," I mean someone who just plays for leisure, who doesn't devote a tremendous amount of time to playing. I knew people in college for whom gaming was a way of life: they would miss sleep to play, they would skip classes to play, and some of them would rather play games online than hang out with people in real life. Those are "hardcore" gamers, to me—the MMORPG players, in particular. Actually, I think the online component is important to hard-core gaming: "hardcore" gamers to me are people whose primary form of socialization is online in these games like WoW, and for whom their sense of identity and social status is linked to their playing ability. (I guess the pre-internet version of this would be people like Billy What's-his-name from that *King of Kong* documentary—somehow being really good at *Donkey Kong* makes this guy important.) I consider my own habits casual because, among other things, my sense of identity isn't at all tied to my gaming ability (which is a good thing—I'm not very good at these games). I just play to amuse myself from time to time, and honestly if a game gets too hard I lose interest—I play to relax, not to be frustrated. I will occasionally devote a lot of time to a particular game because the kinds of games I play are immersive (when I first got *Morrowind*, I would play for up to six hours a day, and that went on for weeks—I just got very involved in the game's "world"), but then I'll go months without ever picking up a controller.

Q: *How have your gaming habits changed over the years?*
A: I suppose my interest in video games began when I was a kid, because I was one of the few kids who didn't have a console (my mother didn't like them—the whole "go outside; those things rot your brain" kind of thing). So when I'd go to a friend's house, and that friend had a Nintendo or whatever, I'd want to play, because it was something I didn't get to do normally. Then it was *Super Mario Bros.* and *Duck Hunt*, and a weird little puzzle game called *Lolo* or something like that.... And when my parents divorced, my sister and I would play video games at my father's girlfriend's house. Eventually my father convinced my mother to let him buy my sister and me a Sega Genesis for Christmas—I was probably twelve or so at the time. Then it was *Sonic the Hedgehog*—by the time we got a system, I think they were up to *Sonic 3*. In high school, I played a decent amount. I was more or less a loner, and I tended to find refuge in repetitive activities. Video games fit into that category, though so did Solitaire and pool on this little game table my dad got my sister. Video games were a distraction—they kept me from feeling lonely.

I didn't really play much through college, mostly because I didn't want to end up like my hardcore gamer acquaintances, though I did occasionally play some of the simpler games on the school servers; trying to blow my friends up in *Super Maze Wars* at 4 a.m. was fun (I didn't stay up to play the game; I played the game because I needed something to do when I was up late), and *Snood* filled my need for repetitive distractions in between classes. But aside from the occasional game on the computer, it wasn't something that took up my time. I didn't get back into it until after my A-exams in grad school: it was the last big thing before the dissertation, and in the summer between finishing the exam and starting work on the diss., I got married and we got ourselves an Xbox as a wedding present to ourselves. That was the summer I played *way* too much *Morrowind* (as did my husband). I suppose that my gaming habits shifted at that point: I went from playing repetitive, simple games to playing big, immersive RPGs. I'd never played RPGs as a kid, but once I played *Morrowind* I was hooked—I liked being able to create and develop a character, and I was endlessly fascinated by this whole other world that I could walk around and act in at will. And I'm the kind of person who enjoys the idea of being able to shoot fire out of my hands, so instead of just a distraction the games became a kind of escapist fantasy for me. Not that I felt a great desire to escape my life—I was happy enough—but it's nice to feel powerful and that's how those games make you feel.

Phone interview with a female player at 66 who lives on a farm in Arkansas.

Q: *How did you first find out about casual games like the games on Gamezebo?*

A: I started playing games when *Pong* first came out on a black-and-white TV. Then I bought a Nintendo and we used to play *Centipede* and those games, and then when I got my computer and Internet that's when I started. I got more into gaming since I retired six years ago. It's really then I've been extensively into gaming, casual games.

Q: *How often do you play?*

A: Every day normally. It all depends upon the weather. Now, this is going to sound crazy, but if it's during lambing season or something happening outside—I don't have the game on the whole day. Usually I will get up at five o'clock in the morning and I turn on the computer to all the stuff that I want to read, and then I will go out and take care of my animals, and then I'll come in and a lot of times, I might sit here and play the game to twelve or one o'clock.

Q: *How much do you think you play on average?*

A: I would say five to six hours on average.

Q: *Why do you play casual games?*

A: To be truthful with you, it's to keep my mind alert. This also why I read.

Q: *What other kind of games do you play these days?*
A: I like the marble games. I just got through a game called *Rock Garden* where you have to line up five pieces. I'm enjoying that game because you have to think ahead. I don't like time management games because those games give you too much to do, and for some reason they stress me out. I don't like shoot 'em up games. I refuse to play them because there's enough violence going on in this world. I don't need to have it on my computer. And there are some games that are really creepy, I can't think of a name right now. For some reason I can't play those and I don't. They get me sick in my stomach. Also I just cannot play any 3-D games or simulations. I don't like those.

Q: *Do you ever play a game you feel is too similar to another game you've tried?*
A: Oh yeah, copycats. I am finding it with those crazy hidden objects games. The quality of those games is really going down because everybody is getting on the bandwagon as we say. There are too many clones.

Q: *Did it also happen to time management or match-3 games?*
A: Yeah.

Q: *So, the same thing happens—if it becomes popular then quality goes down?*
A: Yup. With match-3 games, with the card games. With some of the word games. To me the quality is, you know, like forget it.

Q: *I guess you must also be pretty good at these games by now?*
A: Well, I try to be. But I still cuss a lot when I can't get it. My husband is saying, "Amen."

Q: *I guess you cuss when you are stuck on a level or you're almost getting through a level?*
A: I keep going. Okay, I'll tell you how bad I am. With *Luxor 2*, it came out a couple of years ago. I am still stuck on one level. I cannot get out of it, and I think I have played it about eight hundred times. And it is funny because I keep thinking, "I'm going to get it. I'm going get it." And I go back to it. I don't play it consistently, but I'll go back to it.

Q: *Would a game like that be better if you weren't stuck on it?*
A: No, I keep playing it because hopefully I will figure it out. I never quit a game because I can't do it.

Q: *When is the game too easy and when is the game too hard?*
A: Most of the hidden object games that are coming out, they're too easy. Most of your match-3 games, they're too easy.

Q: *What makes them too easy?*
A: You can go through them in a couple of hours. Some people will say, "Oh, this will be a great game. . . . I finished it in five hours." I don't want a game I can finish in five hours. I want a game that's going to tax my brain.

Interview with a 40-year-old woman who plays online *Scrabble* and crossword puzzles.
Q: *What games do you play?*
A: I play online *Scrabble* and I do crossword puzzles, pretty much.
Q: *How often do you play these games?*
A: Every day.
Q: *How long do you play at a time?*
A: Anywhere from thirty minutes to three hours. Three hours on a bad night.
Q: *A bad night? Why is that a bad night?*
A: Because with online *Scrabble* it gets excessive. You get determined to win and you keep playing and you stay up too late playing. Like a book you can't put down.
Q: *Do you think the games you play such as online* Scrabble *are "computer games" or "video games"?*
A: No.
Q: *Why not?*
A: Because I'm not playing against the game.
Q: *So online* Scrabble *is not a computer game?*
A: No, because I'm playing with other individuals.
Q: *Have you ever played other electronic games, in the arcade or on computers?*
A: When I was a kid I had a ping-pong game, and in junior high school we hung out at the arcade, so I played *Pac-Man*, I played *Space Invaders*, I played *Asteroids*. My father has had lots of video games. He's much more into them than I am. Before I was a teenager really, I was really into games. I really wanted to play them and then I got bored with them in my adolescence.

E-mail interview with a 75-year-old female player of downloadable casual games.
Q: *In your opinion, are casual games (games like the ones on the Gamezebo website) a type of computer/video game or something else?*

A: If "computer games" are a kind of role-playing game on the PC or longer stories like *The Simpsons*, then "casual games" are not computer games. They are rather a type of puzzle. When I was a child, the *Family Journal* [Danish journal] had a page called "Dr. X." It had hidden object pictures, math puzzles, riddles, and more. "Casual games" games seem more similar to pages like that—just more elaborated and refined.

Q: *How did you start playing casual games?*

A: In 2000 I acquired a computer because one of my sons moved to Italy with his wife and child. I wanted to e-mail them. Some friends showed me a computer-based Mahjong game, which I found amusing since we had played it on a beautiful set in my childhood home. Shortly after I found some games in a CD-ROM magazine from the library where I worked. This included Mahjong, *Yahtzee*, and various puzzles. That was how it began. The next game was *QBeez*, which I got completely hooked on, and so on.

Q: *How often do you play?*

A: Almost every day, and often several hours at a time.

Q: *What do you think you get out of playing casual games?*

A: It is a pastime, but it also helps keep my 75-year-old brain in shape.

Q: *Have your playing habits changed over the years?*

A: I have become slower, and some of the hidden object games have graphics that are too detailed and small for me to see without straining my eyes.

Appendix C: Developer Interviews

The following pages are excerpts from game developer interviews conducted between December 2007 and October 2008. The developers represent both downloadable casual games and mimetic interface games. To get a broader representation of the game industry, I also interviewed Frank Lantz, developer of pervasive games, and Warren Spector, well known for developing decidedly non-casual games such as *Deus Ex*.

I conducted the interviews to learn how the developers view the current changes in the game industry as well as their perspectives on the term *casual*, the audience of their games, and the position of casual games in the larger history of video games. The interviewed developers do not always agree with each other or, indeed, with several of the points I have made in this book. I have preserved the disagreements and dissenting voices here.

An Interview with David Amor

David Amor is creative director at game developer Relentless Software, which is best known for publishing the quiz game series *Buzz!*

JJ: It seems you have made a transition in your career. Your CV has games like Quake III: Revolution *and* Space Hulk, *but now you make social games. You have moved quite some way.*

DA: For a couple of reasons. Quiz games seemed new and interesting as a genre and we joined Sony at a time when *EyeToy* had already sold about ten million and *SingStar* was at about six million. So it was clear that if you came out with the right game it could be commercially successful. But also as I was getting older, I wasn't playing the same games that I was when I was 25. I had less time and spent more time playing games socially and less time...in those solo experiences up in the bedroom. I had a family. So I understood why those games are appealing to people.

JJ: So you're the lapsed hardcore gamer, right?

DA: Exactly, right.

JJ: If I said that you made casual games, would you object?

DA: I don't object, but I think casual means different things to different people.

JJ: Do you think it is a useful term? Is casual/hardcore a useful distinction?

DA: I think "casual" can sometimes be a bit too broad. If you think about Solitaire, you can say it is a casual game, but it's some way removed from the games we do. What's common about them is they are for a very wide range of people but we are trying to do things which are social and which you play with other people. In the games we make we try to create most of the entertainment off the screen. There are lots of games that are defined as casual like *Peggle* and Solitaire but I don't think those are quite the games that we are trying to make. I think maybe they are in the same broad category, but *social games* is a term that I use.

JJ: Is there sometimes a big difference between single-player and multiplayer casual?

DA: I think what's common about all of them is that people often think that we are making games for the non-gamers. But we're trying to make games for everybody. I hope the games that we're making aren't excluding the people that play *Grand Theft Auto*.

JJ: I find that in discussing games for a broad market, there seems to be two ways in. One is to talk about game design and one is to talk about players. What do you think is the best starting point? Should I ask you to tell me about the games you design for a broad audience, or should I ask you to tell me about the audience?

DA: Well, I'm less in tune with the demographic that we're selling to. If you ask Sony's marketing department, they can give you lots of information about that. I work from a game design point of view; I think about what features that my sister who doesn't play games will understand. I tend to think of it from a game design point of view.

JJ: Do you see a major change in the industry as such, developing for a new audience?

DA: Well, I think anybody can see that Nintendo has done very well positioning the Wii as a casual machine. You could argue that maybe they appeal to a slightly different demographic even with their Mario and Zelda series, but I think there's no question that this time around with *Wii Sports* and *Wii Fit* they're really going after a different or wider demographic and have been successful in doing so. As for whether there's resistance in the industry, I think that for the most part the industry is made up of traditional core gamers. There is a lot of passion in the industry for making the games that they want to play. As a result, we end up making games that are for our peers or for ourselves. I think most people working on *Wii Fit* would rather be playing Mario.

But it's taken time for the industry to get mature enough to realize that just because you don't like it, it doesn't mean that someone else doesn't like it. I'm quite sure that if you're working on the latest Barbie doll game at Mattel then you're probably not into Barbie dolls, but you understand that they're commercially successful.

JJ: There seems to be a change where people are more likely to design for somebody else than themselves.

DA: I think it all stems from the fact that developers will follow the work, so if the publishers say, "We want you to make this game, please. Here's a ton of money," then people start making those kinds of games. It's led by the publishers and the publishers see success in that market now.

JJ: Is there a danger in this? When you design a game how much of an image of your audience do you have? Are you designing for yourself or do you have the image of an ideal player of your game? What do you know about your audience? You mentioned designing for your sister?

DA: I think it's safe to assume that the people [who] are playing games already are going to be able to play your games. It is different if you aim them at someone who has never played games before. By aiming at the widest possible set of people, and that means my sister, my dad, and my kids, then I can be sure that I'm reaching as many people as possible. The challenge for us is always to make far simpler ways of navigating around menus or moving characters around. Navigating the menus with the Sixaxis controller could be far quicker with less button pushes. But we're not making a game for ourselves.

JJ: I was wondering about one thing. I have been looking a lot at downloadable single-player casual games. One thing that strikes me is that in single-player casual games, you talk about being gentle to the player. But I was just playing Buzz!, and I answered a question wrong, and the whole audience was booing at me.

DA: Sorry about that.

JJ: Do you think there is something different about the role of failure or humiliation when you're playing a social game?

DA: *Buzz!* is really designed as a multiplayer experience. So, what we're trying to do with the host's comments, with the game play, with the crowd's reaction, is to create things for other people in the room to react to. When it delivers a sharp line about how badly you've done, it's not there directed at you. It's almost directed at the other players so that they can ridicule you and you can give them a hard time in return. We're just trying to set up things, situations off the screen. Sometimes people wonder whether or not that works better in Europe and Australia than it does in America and Japan, and I don't know since I'm not from those countries. But certainly in Europe that seems to work, making fun of people.

JJ: Is this is something you think about, how to maximize the value of the social situation?

DA: Always. That has always been the key to *Buzz!*, what we call off-screen interaction. The chance of my video game being able to entertain you and make you laugh more than your friends sitting next to you is slim to none. I am not going to be able to do it because I am not there; I am just a fairly simple computer program that reacts to a set of statistics. But the person sitting next to you knows you, and likes to make fun of you. You've got a lot of history together and so, a much better way of having a good social experience is for me to just find ways of making that happen. I demonstrate *Buzz!* a lot and when I play it with journalists [who] don't know each other, it's not a fun game. As soon as it's a set of people [who] know each other, then it's completely different. It really does rely on a good rapport with people you share a sofa with.

It is all about "How come you knew that question about Van Halen? I didn't know you were a Van Halen fan? Oh, that was when you grew your hair long when you were 16, oh I remember, that was bad, you were going out with so-and-so." I can't ever hope to replicate that, but hopefully the people on your sofa can. That is the idea, to try to bring those kinds of things around.

JJ: As a final historical question, do you see a pressure in the game industry away from making these big hundred-hour games? Do you see that?

DA: I just think that the selection of games is getting wider. I think that *Grand Theft Auto* still sells a lot more than I sell, so you won't see those games going away any time soon. But you will see a wider range than you do now. From traditional games all the way over to *Wii Fit* or *Brain Training*, things that are a long way from what we consider traditional games. And as for the hundred hours of gaming, a colleague of mine looked at some *Half-Life 2: Episode One* statistics, and for a six-hour game, maybe 40 percent or 50 percent of people finished it. So, you have to ask yourself, why are you spending a lot of money on things that most people don't see? Maybe it is a better idea to give them a shorter game. I think it is okay if *Buzz!* gets played six times and then put in a cupboard until next Christmas. I don't think there's anything wrong with that. But, as a result, I'll make sure that I spend my money on delivering it all upfront so you can see all of my *Buzz!* game in half an hour. Hopefully, in that half an hour it's quite a rich and fun experience.

An Interview with Sean Baptiste

Sean Baptiste is the manager of community development at Harmonix, developers of *Guitar Hero* I–II and *Rock Band* 1–2.

JJ: Do you see a major shift in the game industry, a change of focus to a broader audience?
SB: There have certainly been efforts over the years, but people who don't buy video games are a much harder market to reach, but a really big market. I think there are efforts but I think also it's just a natural development of society starting to come to accept video games as a part of culture, if not an art.

People who grew up playing video games are now middle-aged. Video games are no longer this new weird thing. They are starting to become widely accepted. As we get older as an industry, we're finding better ways of interfacing with the game, eliminating the sort of abstraction from what's happening on the screen to some weird plastic device you're holding in your hand.

JJ: It seems like you're talking about what I call the "import strategy of innovation." You are taking things that are on some level familiar to players, and you are basing games around them?
SB: Well that's it. Rob Kay gave this excellent speech at Game Developers Conference about the differences between *Amplitude* and *Guitar Hero* or *Rock Band. Amplitude* was this fantastic game. *Frequency* was this incredible game. They were really good games, but the abstraction of what was happening on the screen in relation to the controller—by looking at

it, you could not figure it out. You actually had to play it to understand what was going on, whereas everybody can understand strapping on a guitar or playing air guitar, or doing drums in the air when you're driving down the street. Everybody understands that. Taking those experiences that were already well known, and making something that is more of a one-to-one process for figuring it out, that is gigantic. That brings in a lot more people. Even for me I feel that the new controllers have way too many buttons. I'm only 30 but I'm starting to feel really old when I look at these controllers.

JJ: There is this feeling that "how many buttons are there on a PS3 controller again"?

SB: It's really quite a bit.

JJ: But the interesting thing is that a lot of people who don't play games report that as a barrier.

SB: Yeah.

JJ: They say, "That's intimidating."

SB: And even for me it's a barrier. And I have grown up playing video games. I think a lot of people will be perfectly fine with it, but as I get older I just want to simplify.

JJ: Some people feel humiliated when they fail, as it stops the songs and ruins it for the entire band. Is that why you have added the new no-fail mode in Rock Band 2?

SB: If you are somebody like me, who has played video games your entire life, for whom it has always been a hobby, then you have always been excited about showing video games to people. But the failure barrier has always been so high. The ability to demonstrate your game to people who have never played before and eliminate that barrier is a very big deal.

At the same time, we have also received some burn back from some of our more hardcore players, who were telling us, "That is not fair, they should learn how to play, life is all about failure!" and so on. And we tell them, "Guys come on, it is a video game, it is a music video game."

JJ: There is a certain hardcore ethic saying that "this is the way a game should be played"?

SB: They don't want their efforts at besting the game to feel diminished by having somebody who has never played it come in and finish a song.

JJ: Another thing that Harmonix does compared to earlier instrument-based music games is that you have added a style layer. Your game doesn't just signal what to do with your fingers but also signals the kind of pose to have while playing the game.

SB: Yes. We have added heightening moments like lifting your guitar up, or, in the case of *Rock Band*, having drum fills or an opportunity for the singer to yell something into the microphone and save a band member [from failing in the game]. Dressing it up with those sorts of game play elements I think enhanced the overall feeling you're supposed to be having when you're actually playing the game.

JJ: I was also thinking in terms of art style.

SB: The art style is something we have worked really hard on for a lot of years. I think a lot of our stuff does have this very distinctive, really interesting look. I am actually really proud of how *Rock Band* looks. We have taken something that is somewhat realistic, but tried to avoid being cartoony, while avoiding the uncanny valley. [We wanted t]o give this slight style to it, with these animations that are realistic but just sometimes a little bit off, to create the world, essentially.

JJ: Do you feel the games that Harmonix makes are an extension of existing games or are they something completely different? Are you reaching back to something like Pong *or are you reaching back to board games? How do you see this?*

SB: It does hark back to those almost pre-video game times when a family would just sit around and play board games or card games. Of course, our games are way cooler. But our games have this classic element of getting together a group of people and having shared goals for entertainment. If you play *Rock Band* like that it's not a competitive game at all. It is entirely cooperative. Even with board games, you are playing against your friends or your family. With this game, it is just collaborative for the most part.

JJ: You can make the case that early video games were games for everybody. Then there was a period of time when video games became incredibly complicated. And now we see a return to simplicity. Does that make sense to you?

SB: I think it does. In the sense that the industry, whether for financial reasons or just for psychological ones, wanted to stretch [its] muscles. So you get to this whole thing where technology limited us, right? There was only so much you could do with all our tape decks and so on. You weren't going to get much more complicated than *Adventure* on the Atari 2600. It simply wasn't there, the graphics weren't there. The technology wasn't there. But there were still these incredible pieces of art that people were coming up with, but it was not possible to make it so complicated that it would only skew to a hardcore audience. But then eventually, as the technology got more advanced, developers just wanted to use it. It is like when the first synthesizers came out: as soon as they came out, every band had a synthesizer. Whether they needed it or not, every band had a synthesizer. Every song had synthesizers.

JJ: And they were playing all the time.

SB: All the time. And it was just, you know, it wasn't a good thing necessarily. But everybody thought, "Oh, here's this new piece of technology. Let's utilize it." I think that is what happened with the industry. After the arcade started going downhill and after the video game crash in the mid-1980s, everybody thought video games would go away. The people who didn't give up were still big fans and their lives were about video games. They would still play video games, they would be the audience, they would buy your games. So I think the game industry *had* to target that audience to save their companies. That whole story of the video game crash probably had a large effect on the video game industry.

An Interview with Daniel Bernstein

Daniel Bernstein is president and CEO of downloadable casual game developer Sandlot Games best known for *Westward, Build-a-lot, Cake Mania,* and *Super Granny.*

JJ: What do you think is the best starting point? Is it "Tell me about casual games" or "Tell me about casual players"?

DB: It has to be about casual players. I think that's a better [way to approach it]. The reason for that is because it really starts with them. And who is the casual player? And I think that definition is likely to be different from company to company and by knowing what the definition is, you can get... at the heart of what exactly makes companies like Sandlot, Big Fish, or PopCap the type of companies that they are. It's how they understand the market.

JJ: Do you feel casual game is a bad term?

DB: I think *casual game* is a term that will get antiquated as more and more people play. Right now, casual games are really an offshoot of the regular games industry. I think there will come a time when casual games are themselves mainstream games and that [the] hardcore games industry is subset of the mainstream game industry.

JJ: Do you feel it's a misleading term in some way?

DB: Not at this point. Probably in about two to three years it will be the majority of people playing casual games and the small minority playing the hardcore games. So I think in two years, a better term would be *main-stream games*.

JJ: That leads to the question: when you design games, do you have a clear image of your audience? To what degree of detail? Do you know what TV shows they watch, or—how do you think about it?

DB: Our audience is people. They usually have two eyes, two arms, and then they have two legs and that's where we end. There, right there. [Those are] pretty much the assumptions that we make about our audience. We don't go any further because I think any assumptions made in the very early stages of design about demographics, whether it's gender or style of play, are superficial and tend to actually cripple you and cripple the design process. I think what we do is to open it up and say that anything is possible. Within this realm of everything that is possible, what is [the] most fun that could be had by the most people? And what has not been done yet that needs to be done? What do we feel good about taking risks on? And we make those decisions on a daily basis across all of our product lines.

JJ: One thing that seems interesting is that a lot of casual game developers are on some level not really part of their own target audience. Do you see yourself as a member of your target audience by virtue of having legs and eyes and so on?

DB: Well, I am certainly. I'm married and have a child on the way so my time is very limited. As a result, I want to have a very good experience with a game, but I simply just don't have the time that I used to have to be able to put into *World of Warcraft* or any large game. I like my entertainment in bite-sized chunks. It's almost like a TV style of entertainment. In the case of *Westward*, I get that by being able to play a mission of *Westward* here and there, save the progress, and come back to it the next day. And I can play for twenty minutes or I can play for six hours and it's still a great experience. In the case of *World of Warcraft* in twenty minutes I wouldn't have been able to install the game, and then I would have to spend the next five hours trying to create my character. It's simply too long of a runway for the type of gaming experience that I'm looking for.

JJ: But do you have an image of the casual game audience as people who are not willing to make very large time commitments?

DB: I think that is a valid assumption. I think they are opportunistic gamers. They are interested in scouring the Internet for the best type of entertainment they can find. It is not necessarily that they are interested in any particular game or that they are looking for a particular type of experience. They are just interested in passing time with a game that they like. If they see something they like that pulls them in, that's what they're going to play, that's what they're going to buy. It is like the difference between TV and film. In film you know what movie you're going to go see. So, you go ahead and go to a movie, buy the ticket, you get popcorn, and you sit there for two hours. There's nothing interrupting you and you watch the movie, and then, you leave and you're done. In our case it's more like TV because it's opportunistic. You get on a website, just like you click through TV channels and you see a game that you like, you download it, you play it, it excites you. It gets monetized in a number of different ways, whether it's through advertising, a $20 purchase, or a subscription. Much like in the case of cable TV. There are different tiers of cable TV monetization. That is where we're at.

JJ: Sandlot Games seems to push the boundaries a bit more than many other downloadable casual game developers.

DB: We always have. Even since the founding of Sandlot Games, when we worked on *Tradewinds* as one of our first projects. Due to the fact that we've been in the industry for a long time, a lot of our franchises are extremely popular and have a following. For example, *Super Granny*, which is a platform game using the keyboard as interface. A no-no in the casual games business, but it was one of our more popular games.

JJ: That's quite interesting because there really are all these ideas of no-nos in the industry, and certainly using anything more than the left mouse button is one of the things we hear all the time. So the question is, how do you deal with selling something new to the casual game audience? Do you have a philosophy on how to do that?

DB: Our philosophy is that "If it hasn't been done, it should be." That is our philosophy and in the case of *Westward*, we read an IGDA report that said that real-time strategy games were the worst games to make for the casual games market. And we said, "Okay, let's go ahead and make it a real-time strategy game." So, my perspective is that it's not about the game type. The question is about presentation, it's about how you build the game and how you present the game to the end-user that makes it casual or not. And so that's where our strength is. It's making games that you wouldn't think are necessarily casual games and making them into really, really strong casual games that lead to casual brands.

JJ: This year at the Game Developers Conference somebody was proposing that in the future, casual games would be the big part of GDC and core games would be a small corner. So, do you see that happening, not necessarily as the

Casual Game Summit taking over the Game Developers Conference, but in terms of the focus of the games industry?

DB: I see the focus of the games industry becoming casual, absolutely. And I think you are starting to see the consumer migration in that direction anyway. So that is going to be the case.

JJ: Let us talk about history. Where do you see the origin of casual games? When did casual games begin?

DB: I think PopCap and Gamehouse were the first ones to really capitalize on casual games. It really began with the dot-com bust. At the time of the dot-com bubble burst, web games were being monetized for advertising as a way of driving traffic on websites and consumer-based websites. When that fell apart, there was interest on the part of the portals to monetize these games in a "try it before you buy it" model. Real Networks started selling a bunch of games. When I was at Monolith we did a deal with Real Networks to take some of our "core" titles, which at that time we thought were going to sell. We took games like *Tex Atomic's Big Bot Battles* and other titles like that into Real Arcade, but the games that were more successful were *Bejeweled* and *Collapse*. That opened the eyes of people, of distributors, in showing that there was money to be made in a completely different type of game dynamic. Not through advertising, but in the "try it before you buy it" model and everything took off from there. What happened with *Bejeweled* and *Collapse* was that PopCap and Gamehouse really scratched the surface of the casual games market. It showed that there is so much latent demand for these kinds of games that it brought other game developers into the fray, like Sandlot.

JJ: How big of a surprise was it that games like this would sell rather than, say, core titles?

DB: It was never a surprise to me because I myself am a casual player. A lot of the producers make fun of my game-playing abilities because I'm very bad in most hardcore titles. But I am good at casual games because that's what I like. I enjoy them. That also guides my design philosophy in looking at games. We can take a hardcore game idea but then it needs to go through the filter of how we casualize it. How do we make it into a title that is going to be truly mass market? So, for me it was never a surprise because it's the things that I like. At the time when I was at Monolith, I worked on a game called *Blood*. These are games that I never played; I never really played *Blood*. I never really liked the first-person shooter genre. I had a little bit of a personal crisis there when I figured out, that when I'm fifty years, is that what I want to be known for, as the guy who built *Blood*? Is that the sum total of my contribution to society and to the world? I think casual games are an opportunity for us to reach a broader base and to really create something of what I believe is a more wholesome type of significance than the games that are within the hardcore games market. That is obviously a very personal decision; everyone is different.

JJ: Do you feel casual games to be an extension of previous games or are they a radical break in game history?

DB: If you look at the history of gaming, games started out being very casual. If you look at early games from Nolan Bushnell's days at Atari, and *Missile Command, Pac-Man*. Those games are very casual and when we were kids in the arcade that's what we played. Somewhere down the line I think the game industry lost those customers. They lost them to other things. Some customers evolved into hardcore gamers. But the great majority, I think, either stopped being gamers or got their entertainment elsewhere.

JJ: That's one of the things I realized lately. I always ask people if they play video games and they will say, "No, I don't play video games," but it will usually turn out that they have played Tetris *or* Pac-Man *and they really liked it.*

DB: That's the thing, historically even hardcore gamers started out being casual. There was no such understanding back then. And you know we're starting to tap into the market of gamers that have never played games; as a result of course, they're starting out with the casual titles just like we did. Or folks that have not played games for a long time are returning to that experience and they're having a little bit of nostalgia about it.

JJ: I have this question written down here, which says, "In 1977, all video games were casual, agree or disagree?"

DB: Oh, absolutely.

An Interview with Jacques Exertier

Jacques Exertier is a game developer at Ubisoft Montpellier. He was the cinematic director of the Wii game *Rayman Raving Rabbids*.

JJ: If I said that with Rayman Raving Rabbids *you made a "casual game," would you object? Would you prefer a term like social games, or another term?*

JE: First, let me just tell you a quick anecdote. I looked up the exact meaning of casual in a dictionary (I am French). I was surprised to find different meanings like: Unconcerned (a casual attitude), irregular (casual workers), not committed (casual sex), by chance (a casual observer). I hope *Rayman Raving Rabbids* wasn't an "unconcerned irregular not-committed and by-chance game."

We can speak of casual, accessible, and social games. Each of these terms means different things.

- *Casual* refers to the player's state of mind—the player's commitment in the game. This is a relation that goes from the player to the game.
- *Accessible* refers to the game itself—that it is easily played. This is a relation that goes from the game to the player.
- *Social* refers to people (players or observers)—the relation between them generated by the game. There are relations on a level about the individual player and the game.

We tried to keep these three spheres in mind when designing RRR.

JJ: Do you find casual/hardcore to be a useful distinction?
JE: Let us imagine two game development teams. The first one has designed a "casual game"; the second one has designed a "hardcore game." Their games are reviewed by the boss and he says (two months before the release date), "Team one, I want you to target the hardcore audience too." He tells team two, "Team two, I want you to target the casual audience too."

I feel that the challenge for the first team is hard, but possible without losing their original concept. But I feel it is almost impossible for the second team without completely changing their concept.

JJ: Do you see a shift in the industry from what we could term hardcore to casual games and players? If so, when do you think this began?
JE: There is definitely a shift in the industry, but it is more than a corporate decision to reach a new audience. Above all, I think it is because the first generation of gamers who were teenagers at the beginning of the history of video games have now grown up and become adult gamers. Most of the large adult-casual audience today is probably composed of this type of gamers. Of course every business wants to reach a large audience, but the possibilities for doing so have changed.

If the Wii had been launched twenty years ago, would it have met the same success? I am not so sure. In ten or fifteen years, we will probably observe a new shift in the industry toward senior citizens because it will become possible to target the first-generation gamers who will then have turned sixty.

JJ: *When designing* Rayman Raving Rabbids, *did you have any design principles for making the game more social, or for utilizing the existing social relations between players?*

JE: The main design principle we use was to playtest and iterate, and we focused on what happened between the players. Social games have to get the game outside the screen.

In *Rayman Raving Rabbids* there is a minigame in two phases: In the first phase, you have to run as fast you can by shaking your controllers; in the second phase you have to stay still and point into a small circle slowly moving slowly on the screen.

When we were playtesting this, we observed that some players or onlookers began pushing the person next to them to make them miss the circle. The game had crossed from one side of the screen to another. Even onlookers were playing the game, not only observing and laughing but interacting with the game without even touching the controllers. That is a good design goal.

JJ: *Do you have design principles for making your games accessible to gamer audiences as well as an audience not used to playing video games? What are the pitfalls?*

JE: I think casual gamers are less involved in competition than the hardcore. They particularly need another dimension in the game to be hooked (it can be humor, learning something, playing with others). We put competition in the *Rayman Raving Rabbids* minigames (with the score mode) to offer challenges to the more hardcore gamers. But I'm not sure that was the key point that allowed us to reach them. Of course hardcore gamers are not only interested in competition. The playtests showed us that both casual and hardcore gamers were laughing and having fun at the same moments for the same reason. And that was not particularly when they were winning or when the challenges were harder.

I think you have at least two approaches to reach the two audiences at the same time:

1. Design casual challenges and add depth and difficulty for the hardcore gamers (the solution I was thinking of [for] team one in the previous question). This is probably not the best solution because in a certain way this is two different games targeting different people at different moments with different means.

2. Find and develop other dimensions, more universal and not only victory-based. Try to make casual and hardcore players meet each other on the same emotions with a social game.

The strength of *Rayman Raving Rabbids* is humor. On-screen, the rabbits are stupid. Off-screen, players are moving and shaking their Wii remote like mad people. I think that works well because of this similarity and universality between the stupid acting of the rabbits on screen and the stupid moves of the players off screen. This works on both casual and hardcore gamers. Perhaps social games manage to erase the gap between casual and hardcore.

An Interview with Nick Fortugno

Nick Fortugno is a game designer who previously worked at the company Gamelab, where he was lead designer on titles including *Diner Dash*, *Plantasia*, and *Ayiti: The Cost of Life*. He has now cofounded a new studio called Rebel Monkey.

JJ: What do you think is the best starting point? Is it "Tell me about casual games" or "Tell me about casual players"?
NF: I think the *casual game* term is fraught at this moment. We can define casual games by the audience. If we define it by the audience then it's games for 30- to 50-year-old women predominantly, although it's proven to be a slightly more equal gender split. If we look at it from the games perspective, though, it is a series of design philosophies. I think the safer question is actually asking about the games, because I think the audience is a phenomenon of the business model more than anything else and that when we start to see the proliferation of casual games into other channels, we start to see that other people will play them besides that audience.
JJ: But the distribution channel and the audience and the business model are mutually dependent, aren't they?
NF: Yes, but the thing to remember is that when the casual audience appeared, it was a surprise to the people making casual games. No one thought there would be games for 40-year-old women. That is fascinating to me and it implies an emerging audience from all these other factors.
JJ: OK, but what is a casual game and do you think casual games is a good or bad term?

NF: I don't love the term *casual games* but I think it now is an inescapable term. I think that data shows that the term *casual* is not exactly accurate. But when I ask myself what casual games are, the answer I come up with is a set of design parameters that determine how a game is made. It has to do with the intuitiveness of the core controls; the use of failure or progress as a reward structure for continued play; the steepness of the learning curve, and the use of certain kinds of content as motivating factors. I think those are the things that determine something as a casual game or not a casual game.

JJ: But then tell me about the points you just mentioned, such as intuitiveness. What is a casual game, then?

NF: A casual game is a game with a simple set of basic controls that then becomes complex through context. Specifically, level design and the development of level design. So, in hardcore games, if you pick a game like *Civilization*, the interface is very complex. It is complex but its purpose is of allowing you a kind of micromanagement control over what you are doing. Similarly you can point to a game like *Call of Duty* and say it has a similar kind of thing, where a complex set of controls allows you a maximum amount of actions within the game space. Casual games move on a different approach, which is to take a very simple interface that allows easy access to the game's system and then makes the complications arise from the content in which you use that control scheme. That's one major difference. The second major difference is that hardcore games will teach you through failure. So, a hardcore game will teach you by having you fail multiple times and retry, which implies a threshold or tolerance for failure that's fairly high. A casual game goes through the same learning process instead by using an incremental reward structure that allows you to succeed your way while you slowly learn lessons of play. I think that the casual game player or the people who don't typically play hardcore games are not accustomed to that kind of failure structure and that's actually one of the biggest barriers to accessing games.

JJ: In behaviorist terms you are saying that hardcore games are about negative reinforcement and casual games about positive reinforcement.

NF: Yes.

JJ: So what about the learning curve and the reward structure?

NF: Part of the idea of this with positive reinforcement is that the learning curve is much smoother. The levels tend to be teaching rather than restrictive, the way they are in hardcore games. So, in a hardcore game, think about a game like *Virtua Fighter*: if you play the single-player mode, you don't learn from fighting the first fighter how to beat the second fighter. You just beat the first fighter, then you beat the second fighter. And you get better because you have to. That is not like a casual game. I think *Portal* is a casual game. *Portal* teaches you what's going to happen in the next level. So if you play *Portal*, you'll see that the thing that shoots missiles, that you need to win the boss battle, appears several sequences before the boss battle and the thing you do to beat the boss, you've already done.

JJ: But what you're describing is really a typical puzzle game structure.

NF: Yeah, I think so. And it's not unique to casual games. I mean *Zelda* has been using a structure like that since its inception. But I think it's the reason why *Zelda* is more accessible to non-game players than other kinds of games.

JJ: On that subject, you also once talked about how Diner Dash *does become quite hard, that you are pretty likely to fail.*

NF: Yes, but there is a difference. The question here is not about challenge. Casual games can get very hard, like *Zuma*—very few people win *Zuma*, very few people win *Diner Dash*. That's not the relevant factor. Casual games can, in fact, get as hard as hardcore games. The difference is the way you get to that difficult point: in casual games it is a slow process that prepares you for that difficulty. It may not have been totally successful because you may not finish *Diner Dash* or *Zuma*, but at least you got those levels at a very smooth slope. Whereas a hardcore game—so I'm picking an unusual hardcore game example, just for the purpose of demonstration, like *Katamari Damacy*: for every level you play, you have to learn from the beginning, and you fail on that level. Casual games are more inviting and they hold you by the hand, whereas hardcore games drop you in the deep end and expect you to swim.

JJ: If we switch to video game history and I say, "In 1977, all video games were casual," do you agree or disagree?

NF: Disagree.

JJ: On what grounds?

NF: Difficulty. That a lot of games were really, really, really, really hard, unforgivingly hard.

JJ: Including arcade games?

NF: Even more so, because they were trying to eat your quarters. Think about a game like *Asteroids*. *Asteroids* is extremely unforgiving. There's a button on it that teleports you to a random part of the screen where you could instantaneously die, which is pretty unforgiving. I think that in essence, they are not casual. I am unusual for thinking this way but I think they resemble casual games in that they are puzzle-y and low resolution, and that they involve fairly simple controls. But the place where they fail, which is actually I think the most important place, is that they punish you repeatedly. This is the way of motivating the player that has ghettoized the game industry. The escape from that type of motivation is what allows games to flourish in the larger market.

JJ: I see what you mean. On the other hand I always ask people if they play video games. Some people will say "no, no." Upon further questioning, it often turns out that they actually did play Pac-Man *and enjoyed it.*

NF: Okay. That's another possible definition of casual games. A definition that a casual player is someone who doesn't think of themselves as a gamer. I think that's true too but if we compare it to movies, nobody talks about themselves as movie watchers, because they understand movies as just part of a larger culture.

Look at a film scholar or a movie aficionado, right? I think a game like *Pac-Man* or early Nintendo games fit there. People say, "Oh yeah I was growing up when there were arcades and that doesn't make me a gamer. But I did play *Pac-Man.*" It is a good time for games because of their increasing cultural acceptance. People don't think of games in isolation from the rest of their cultural experiences.

JJ: If you go to a bar, especially in New York, they often have a Pac-Man/ Galaga *machine running. And they are punishing but the controls are also quite intuitive and they require quite little knowledge of video game conventions because they were made at the time when there weren't that many video game conventions. Aren't they casual in that sense?*

NF: There is a spectrum obviously. It's not binary. But they don't have the long-term play success that a game like *Bejeweled* does and I think that's because of difficulty. But I wouldn't say they're not casual.

JJ: But they do touch an audience that is also normally untouched by modern-day hardcore games?

NF: Absolutely and I think it's exactly to the point of the simple interface, right. I think because it's acceptable, right. So, then acceptability is really critical. If we want games to grow, they must inevitably grow into a medium like movies where everybody plays games. These are the two key things that have to be overcome: the complexity of the interface and difficulty barriers that games present. And I think early arcade games for many, many reasons, including simple technical limitations, had to be something really simple, with a joystick and a button.

JJ: When you design games, do you have a concrete idea of your audience? Is it someone you know, or do you have an idea of what TV shows they watch, or where they shop?

NF: To an extent. You know my business partner is very, very pop culture sensitive. She has kind of a sixth sense about pop culture and popularity. So, she'll gravitate toward things that she thinks are successful and draw from them. It is not just that we take initiative to do things that we think people in the audience will like, but that we pull away from things that we know they won't like. We'll avoid violence because we know that's something that is not interesting to the audience. For me that is a set of design parameters. I think of the audience as a cluster of design expectations that I'm trying to map to.

JJ: You are not a member of your own target audience?

NF: Not exactly. Because I have much higher tolerance for hardcore games. So that already puts me outside of the audience. I don't play games as a simple time-killing escapism, so a lot of types of casual games that are successful don't really interest me that much, but then I tend to not to make them, frankly. When I make casual games, I don't make *Bejeweled* and I don't think that's an accident. I'm not interested in games like that, but I can play *Diner Dash* or *Plantasia*, which are more complicated. But on the other hand, I like casual games. I play *Zuma* a lot and I play *Break Quest* a lot. So, I'm not outside of the audience either.

An Interview with Frank Lantz

Frank Lantz is creative director of the company area/code. area/code has made pervasive games, which often take place in urban environments and are played via multiple media channels. area/code has now moved into Facebook games with *Parking Wars*. Before working at area/code, Lantz was director of game design at Gamelab.

JJ: I find it interesting that your company area/code is known for making games on a very large scale, pervasive games. And now it seems that with the Facebook game Parking Wars *you are moving in another direction toward*

making games that ask players for only a very small time investment—this
appears to be the exact opposite of the pervasive games or alternate reality
games that you have worked on previously. Is this a departure in your work,
or is it a continuation?

FL: It is similar in that [*Parking Wars*] is not a conventional video game.
I think that connects these things together. We've been conditioned to
think of video games as a very narrow subcategory of the possible kinds
of games. There has been a tendency to see video games as computer
software where you control a little character in a 3-D environment, and
all the different variations of that. I just think of games in a much broader
sense; I see video games and computer games within the larger context of
games in general. With games and sports, to me there is a big spectrum
of these things and I am interested in them all, and I like them all. That
is one aspect of it. Trying to reclaim this broader palette.

Another aspect of it is a focus at area/code on the interesting ways in
which games overlap with the real world. Where games insert themselves
into your life and the kind of surprising ways that they can overstep the
boundaries, and how that can lead to interesting new kinds of game experiences. In the case of urban gaming, it's about appropriating public
space and rethinking how you turn a familiar space into a game space.
In the case of *Parking Wars*, I think it has a little bit to do with time, you
know that... this is a game that you're always playing, and the rhythm of
your ordinary life becomes part of the game dynamics. When you are
asleep at night, you know people who are parking on your street and
there's nothing you can do. Now, if you are away on vacation and you
don't have access to a computer and your friends know that, then that
becomes part of the game, you know that becomes part of the tactics and
strategy of the game.

JJ: One of the things that seem to be happening is that the game industry is
gaining a slightly more sophisticated idea of the audience. There seems to be a
shift to making games for a casual audience, and that casual audience is
assumed to be nonidentical with the people who developed those games. How
do you see that shift? Do you see a major seismic shift in the game industry
from hardcore to casual or games for a broader audience, or is it more a rhetor-
ical thing?

FL: I don't think that there's a really super-dramatic and profound shift.
I think this is just evolution. I think it's just this kind of glacial change, a
kind of climate change that occurs on a large scale. Being so very big, it's
a very big category of culture. I don't think things like that change over-

night. I think they change slowly. And I also think that it maybe has more to do with how we frame the question, how we frame the category of games. Games have never really been for a narrow audience, though I think games have always been for a narrower audience than story. However, it is still pretty universal and almost everyone plays games, it's just that not everyone plays *video* games. I think part of what will happen is that the category of video games will stop feeling quite so well defined. I think that was always part of my thinking in doing large-scale real-world games, games with augmented reality, alternate reality, and cross-media. These are games that use computers, they have computers in them, but they are not computer games because who cares?

I'm primarily interested in games. And if they're software, that's great! And if they're not software, that's also great. I think that's probably also true of people at large and it's partly what you are seeing with people who are playing *Guitar Hero*: they are less focused on this as a computer game or a video game or as a piece of software, and it's more, "No, this is an activity. This is something that happens to be on a video game system, but it's a fun activity that I do with my friends."

In that sense it's like playing poker, tennis, or golf. If you broaden your view and look at how games as a larger category are a part of people's lives; [for example,] someone grows up and plays chess on the weekends with his dad, then he plays chess occasionally with his friends, then he plays chess with his children. But chess is also a hardcore game: there are some people that devote their lives to chess and learning it and mastering it and becoming professional chess players and going to tournaments. There's a single game there, but there's a spectrum of ways to engage with it and I think that is more typical of the way that games fit into people's lives.

JJ: *I was trying to frame that by saying that part of what's happening with something like* Guitar Hero *isn't that video games are becoming cool, but that they are becoming normal.*

FL: I think that too. I think they are becoming normal and it is also that we are becoming less focused on their status as a piece of software. *Guitar Hero* and the Wii are both objects that you play with, right? It is not just a program that you interact with. It's a game that's made up of software and objects and I think that's an interesting evolution in a way we think about these things.

JJ: *Steve Meretzky has argued that early video games in the arcade started out as being mass market, and then there was a period of time from perhaps*

1981 to the point where Solitaire appears in Windows when basically the mass market was entirely forgotten.

FL: I don't know. I mean it's weird. The thing about casual games as a category, or the casualization of games is that it implies a kind of dumbing down. It implies games that are less complex, less deep, less challenging, and that is not what I'm interested in.

JJ: Does it? That's subject to discussion isn't it?

FL: Don't you think it does? Okay, here's a game for people who don't really like games. So we are going to make it real easy, it's more of an activity. It's a game-like activity, something players can click on and it's kind of hypnotic and it is maybe a little mindless, and it overly rewards people, like in *Peggle*. By the way, I don't think that necessarily holds for *Peggle*. But there is the Jonathan Blow style of criticism of games that overly reward you. You don't really do anything, and [the game] is showering you with rewards. In that sense it's tending to lesser instincts and a kind of dumbing down.

There's another way to think of it: game designers are very ambitious, you have to be ambitious to be a game designer. So what you really want to do is for a living is to reach into people's heads and toy with their brain. You want to create a world that people step into. You are a conductor of groups of people in dynamic social situations. Game designers are very ambitious people and yet so few contemporary game designers have the kind of ambition where they step back and look at a game like *Monopoly*. Or a game like poker. Look at the game of poker, look at the game of football, look at the game of baseball, and look at the game of basketball. These are games that people have architected cities around. These are games that people have played for decades or for hundreds of years. Look at chess. Why shouldn't video games aspire to [the] same kind of status? Why shouldn't the very best of them aspire to have that kind of impact on the world? A game could be something that is worth devoting your life to, for a player to grow up playing it, to spend their whole life playing it... to make it the centerpiece of their entire life.

Is that casual? No, but that's what golf is, that's what baseball is, that's what poker is, and those are the games that appeal to a broad audience. So to me, it's foolish to not look at this larger audience. I think we should think about this larger audience in historical terms. It's not just about reaching a million soccer moms, it is about making a game that can stand the test of time.

I think games, even more than movies or books, have the capacity to cross generations. A game that I learn, and then I love, and then I teach my son how to play and then we play together and then he teaches his son. When I taught my son how to play Go, it became a part of our relationship. This is the kind of ambition that game designers could have. Obviously, there is also disposable culture too; I love pop culture and disposable culture. I love sketches and small things that are designed for the moment, but why isn't there more of a sense of possibility, that a game can be more than just seasonal software? And that's what hobbyist games are: seasonal software. Games that are on the shelves a month and then we move on to the next one.

JJ: I guess there is a limited amount of video games that can have the status of baseball or football.

FL: But some do, like *StarCraft. StarCraft* clearly does, *Warcraft* [too]. I think to a certain degree, you could say that *Counter-Strike* has done that a little bit. It almost has the status of baseball. And are these casual games? No they are definitely not.

JJ: You object to the term casual game?

FL: I don't object. I myself am not interested in that as a way of defining my own work. It is not how we think of the things that we do at area/code. But we do know that we are making games that don't fit within the existing context of hardcore mainstream video game development.

JJ: But let me ask you, what does casual mean to you?

FL: It depends on who is using it. When someone says "I am working on a casual game," it is descriptive only by exclusion. When you say "casual game," the only thing you're saying is that it's not a first-person shooter, a real-time strategy game, a driving game, a high-end 3-D game of some type. You could be making a casual MMO, you could be making a casual puzzle game, you could be making a casual game that's very narrative like an interactive romance novel. There's very little that formally links an interactive romance novel with a match-3 kind of logic puzzle.

JJ: Isn't there a link in terms of the kinds of engagement or a link in terms of the imagined casual player who plays a casual game?

FL: I think it's much more about the audience . . . than it is about the type of engagement. Let's take *Tetris* as an example. *Tetris* might be the *ur*-casual game. If you think of casual games as the PopCap-style match-3 puzzle games, *Tetris* is the blueprint for that, and yet it is possible to play *Tetris* in drunken binges, you are addicted to this activity, this repetitious thing you can't walk away from for hours. When you finally put it down, you are groggy and have a headache. Or it is possible to play *Tetris* when you are standing in line at the DMV and you think, "Okay, I am bored, I've got five minutes to fill and I will play some *Tetris*." It is still *Tetris* in either sense. Is one of those more the casual game and one of them not? I think that many good games are capable of fulfilling both of these roles. So it seems like it's not really a property of the game, or even a property of the game experience that makes it a casual game, it's really more related to marketing or social context.

JJ: Okay. I think it's possible to make the argument that there is something called casual game design, like in Tetris, *which is really about being open to being used in a wide variety of ways. That the question isn't, can you only play it in a very shallow way, but rather, can you play it for a very brief period of time or can you play it for weeks? Whereas traditional hardcore game design does not allow this very short play time.*

FL: I think that's not a bad description.

JJ: I want to ask you about Parking Wars. *That's not a casual game I guess?*

FL: Is *Parking Wars* a casual game? I'd never refer to it as a casual game. I would say that it is a social game; that is a tag I feel more comfortable putting on it. But what does that mean? Obviously, *Quake* is a social game and *Werewolf* is a social game, and *Diplomacy* is a social game. I would say social game is more accurate for me; it's really all about people.

An Interview with Garrett Link

Garrett Link is general manager of social gaming at Real Networks, one of the largest publishers and portals of downloadable casual games.

JJ: Let me start with a meta-question. Which is the better starting point: "Tell me about casual games" or "Tell me about casual players"?

GL: Well, I think casual games are at a very interesting time right now. People define it differently. Some define it as games for a specific demographic, which is certainly focused on the players. Some define it as a game-playing mechanism. And some define it as characteristics of a game-play experience. I tend to find myself in the last two categories. Casual games are easy to get into, take a lifetime to master. They are simple to understand and provide a lot of entertainment. Those are the key characteristics for me and they could be on any platform.

JJ: Where do you see the beginning of casual games? Do you see casual games as an extension of older games or are they something new?

GL: They are definitely an extension of older games. You can go right back to *Pong*. It has all of the defining characteristics of casual in my mind and it is a multiuser game which is interesting for the casual space today.

A lot of coin-up games were casual. They were just designed to take a lot of coins out of your pocket. They were designed to make you fail. It's just an evolution, there's a timeline and you look at why audiences play games online today, they play to relax, to transform their mood, it's a little bit of escapism. Games have broadened out into different things, there is *Second Life*, different experiences.

JJ: When you are working on project, do you have an image of the audience? Are you designing for yourself? Do you have an idea of an ideal player for the game you're working on?

GL: This is different from studio to studio and different from person to person. But let me give you an example because we design as a philosophy with our audience in mind. In our Eindhoven office in the Netherlands, they have actually named the people and their names are Sophie and Marie and there are characteristics about Sophie and Marie that everybody in the office can describe to you: how old they are, how often they play, where they play, when they play what kind of things they like to do, their personas. In all of their game design documentation and in all of their conversations, you'll hear references to Sophie and Marie wanting things.

JJ: Oh, that's interesting.

GL: When I worked in the core space, I built games I wanted to play. And the truth is that I actually enjoy playing quite a few of the games that we build today as well, and for different reasons. I'm married and have family. I have less time, so just that alone makes some of these games appealing.

JJ: There seems to be a shift from the development of core games where there is some overlap between the developers and the targeted audience, and now when you have Sophie and Marie, it doesn't sound like you are in the target audience.

GL: Certainly not.

JJ: But you also play the games, right? So, is there an issue or a special trick to designing things for which you are not in the target audience?

GL: It's a trade, it's a skill, it's something that you learn through experience just like anything else, being a carpenter or whatever. In Seattle, we do have people that would meet the Sophie and Marie criteria. We try to do that, then we use external tools, we focus-test, we get benchmarks against people that we do believe to be the target audience.

But you do have to think about it. And once you think about it as a conscious effort, then it becomes reasonably easy to apply your trade. It is one of the fun challenges, actually.

JJ: But do you think there is a certain danger of ending up talking down to the audience, thinking that "here we are, these cool geeky people, and over there is the weird target audience." Is there a danger in pigeonholing the target audience?

GL: I would say there's a danger in it, but I haven't really seen that happen in practice. People are very proud of the work that they do and they're proud to be serving a segment of gamers that largely, people feel weren't served before.

There are times when we sit in a room and we have a design conversation where someone says, "Oh, this would be so cool," or "Oh, it would be wonderful." Then we ask ourselves, "Will Sophie and Marie like it?" and we have to say "No." It's not in a disrespectful way. I think that's quite natural. Especially when we have passionate people that consider themselves gamers.

JJ: It is strange with casual games, how the developers and the target audience are so distinct.

GL: I think when you talk about the audience that's buying downloadable casual games, in broad terms they are 45-plus-year-old females. There are not a lot of skilled artists and programmers in that age bracket in the digital era. I think over time that'll change. People like me were just out of college before we even had an e-mail address. I didn't grow up with the Internet or Google or any of these kinds of things. If you think of the audience now, they are fifteen or twenty years older than me. It must be even more foreign to them. I think that's in part what's going on.

JJ: It seems like casual games are growing their own sets of conventions. So, is there a way in which casual games eventually become non-casual? Is there a chance of casual games becoming so complicated that they are becoming a new hardcore?

GL: I think games are becoming games again. You're not going to have this hardcore and this casual category. They could be largely just known as games. I keep close ties with a lot of people in the core space and we were having a conversation with one guy the other day who said, "Man, you guys in casual really got it, don't punish the player, help them along," and he was giving some examples in *BioShock* where you respawn in this protective bubble rather than having to restart the level. There are lessons from what's been done in the casual space that are transferable to all games and I think vice versa. I really believe that we are getting back to all games as one single thing.

JJ: So eventually the term casual games goes away?

GL: I believe so. I don't know how long it will take. The distinction between casual and hardcore is definitely blurring and will continue to blur, so at what point do those terms kind of go away, I don't know.

An Interview with Dave Rohrl

Dave Rohrl is director of game design at Zynga. Until 2005, Rohrl was the general manager of PopCap, developer of *Bejeweled*, *Zuma*, *Peggle*, and other successful downloadable casual game titles.

JJ: What do you think is the best starting point? Is it "Tell me about casual games" or "Tell me about casual players"?

DR: Casual games are games that are intended for people for whom gaming is not a primary area of interest. People who don't think of themselves as gamers, people who aren't structuring their time as gamers. When you look at casual games today, they are defined by casual gamers but reciprocally; casual gamers are really defined by casual games. The two are becoming increasingly intertwined. Five years ago, if you asked who is a casual gamer, it was a soccer mom. A 42-year-old, middle-American woman. If you ask who is a casual gamer today; it's somebody who's looking for cheap, accessible, family-friendly, interactive entertainment. You know, in short bursts.

JJ: Is this about the industry having a better understanding of the audience or is it the audience shifting?

DR: That's a really good question. I mean, I think the audience isn't shifting per se, it's broadening. And I think that there are a couple of processes at work. I think the industry learned pretty clearly that the "try and buy $19.99 download" is really good at monetizing one demographic segment. And it didn't have that much appeal beyond that demographic segment. I think it is business innovations that have allowed the rest of the casual game-creation community to reach out to other segments. Most of those innovations have really come from overseas, especially from China.

JJ: Where do you see the beginning of casual games? Will you say "2001, when Bejeweled came out"?

DR: *Pong. SpaceWar!* is kind of hardcore. *Pong* is kind of casual. Simple controls, simple concept, simply visualized, easy to get your head around. You can walk up to a *Pong* machine and you know what to do with it.

JJ: OK. So here is a question: In 1977, all video games were casual, agree or disagree?

DR: *Space Invaders,* I think, is fairly casual. But *Defender* for me is pretty hardcore. There are things that they did differently from what we think of as casual in the modern era. A lot of the modern casual games have to do with a shallow difficulty ramp and engendering a player's feeling of success. In 1977 the idea was really to give a player just enough success to motivate them to put in another quarter and then kill them off.

JJ: Thinking about the audience: when you work on a game project, do you have an image of your audience? Is it someone you know, or is there some sort of ideal player for a game? Do you know what television shows they watch?

DR: I wish I could say I always did that. I think it's a good practice. I often don't. When I'm working on a game, I tend to have a very personal relationship to that piece of work and I'm looking at my own reaction to it on many levels. But then I play a lot hardcore games as well as a lot of casual games. And as I'm reacting with pleasure to a part of the game, I'm trying to figure out which part of my brain is reacting to that and tap into my own inner casual gamer. I love to sit down for a good game of *Word Wonks* as much as the next guy. But as opposed to visualizing that soccer mom or that teenage girl, I'm really just trying to be in touch with my own inner casual gamer.

JJ: Okay. So, in a way you are member of your own target audience?

DR: Yes. If you don't like eating chocolate, you probably shouldn't be making chocolate. A great way to build a bad casual game is have it built by people who really aren't that excited about casual games. You are going to get a soulless mercenary product.

JJ: Some people tend to talk about bling or polish or juiciness, excessive feed-back. What is your take on that?

DR: That's not what I think of as polish. To me what polish is about, you know, the user is going to have hundreds of thousands, millions of individual moment-to-moment interactions with your game; polish is the effort that you put into making those interactions feel better. And it's got value in that respect. It can help that steady drip, drip, drip of neurotransmitters. I always like to leave room for polish, and rough games are rough games. You see them, you know them. They're not hard to spot. So, it's nice when your audience doesn't have to forgive anything.

JJ: But it is also about the communication of in-game events. That it has to be pleasurable on a basic level.

DR: Absolutely. People play games to feel good. People get to feel good in different ways but nobody play games to be humiliated. No one plays games to be bored, we can get enough of that in daily life. People play games to feel good. So if the game is doing a good job in telling them that they're a wonderful person, it helps. Like I said in my talk in Seattle, I play *Bejeweled* because *Bejeweled* says "Excellent!" to me.

JJ: I thought that was really interesting. Because there are different types of effects, there are very purely communicative effects and then there's something else. My father played Peggle *the other day, and he said, "Oh, it's like somebody is praising me all the time."*

DR: Exactly.

JJ: How do you see the future of casual games?

DR: One of the interesting things to me about the casual games category that really differentiates it from boxed products, whether it is board games or whether it is a console game, is that it is an open-box category. On a certain level you still want some flashy packaging to attract the customer's attention and drive interest. But the nice thing about a casual game, in almost any of its forms, is that pretty much everybody has some form of free play available. The customer gets to make a really deep, detailed, informed, reasonable decision about whether or not this game is worth their money.

Casual games basically started out to reach an underserved segment of the market. So, you essentially have the core games industry servicing young males and a firm belief on the part of the industry that you couldn't make money servicing anybody else. And I think that through the work of innovators like Pogo, and PopCap, and Big Fish, and Real Arcade, that perception has been changed forever. Now nobody is saying you can't make money selling games to middle-aged women.

We are continuing to serve that segment very well and I think we're going to continue to see really simple games break out. What's happening now is that we're seeing more casual stuff in really hardcore channels. We're seeing more hardcore stuff in really casual channels and what that hopefully means is that we can create enough variety of products as a collective industry that serves the full spectrum of consumers.

An Interview with Warren Spector

Warren Spector is a game developer famous for working on classic "hardcore" titles such as *System Shock, Thief,* and *Deus Ex.* Spector is now creative director of a new studio called Junction Point.

JJ: Do you think there is a shift from what you could call hardcore to casual in the industry, and to what extent is it a change in rhetoric and to what extent is it a concrete thing happening?
WS: There has definitely been a change even over the last three to five years, and I think it is a substantive change, not just a rhetorical one. The audience is growing dramatically and the economics of even hardcore gaming are demanding that we cast a wider net. The market for hardcore games may not be getting smaller, but the cost of reaching that hardcore audience is basically the same as reaching the broader audience that now exists for games, and therefore it is harder and harder to find people willing to fund games that only go after that narrow hardcore audience. It is a real change, no question about it. It costs the same amount of money to reach 250,000 hardcore real-time strategy gamers as it does to reach potentially five million console gamers.
JJ: I used the terms casual and hardcore. Are those useful terms?
WS: I think there is a useful distinction to be made but as with all other high-level categorizations, the borderlines are dangerous. Someone can get onto Yahoo, Shockwave, or download something from Xbox Live Arcade that is a casual game, but that is very different from my mom getting online to playing Hearts. Casual is the extreme, hardcore is the extreme, and there is a continuum in between.
JJ: Has your relation to the game audience changed since you developed games ten to fifteen years ago and now? Do you have more of an idea of your audience?
WS: The hardcore people I used to work with immediately equated reaching a larger audience with dumbing down, and thought that trying to make a more accessible user interface was a betrayal. And you had to say, "Guys, look at the gameplay", and tell them, "We are just making the larger audience experience what you have been experiencing all along."
JJ: How do you see it, making games for an audience you are not part of?

WS: How do I reach a 13-year-old? How do I reach the bottom of my demographic? There are ways of dealing with this. Trust your instinct. I can't really count on the people who are playing this game to have a historical reference point for what I am doing, so I better train them before throwing them into the deep water.... Finally, you blind test. I am a blind-testing junkie.

JJ: *There was a story discussed the other day that* System Shock *could have done with more blind testing.*

WS: I think that *System Shock* is a work of art, but the user interface had such a steep learning curve and it demanded such skill to use that it just scared people away, and that is unfortunate. That is my point to the hardcore gamers: I guess there is a prestige thing associated with mastering things that other people can't, but I don't want to that, I want to reach those people who can't master the intricacies of an overly complex interface.

JJ: *Isn't that also an industry-wide trend, the concept of blind testing?*

WS: Sure, the concept of blind testing is gradually making its way through the game industry, like the need to reach a larger audience. The audience is getting broader and deeper, game development is getting broader and deeper too. There are opportunities now that did not exist five years ago. You can be four guys in a garage and make a really hardcore game, and there is a way to reach an audience and monetize that investment of time and passion and love. The fact that I am not making games just for hardcore gamers is irrelevant as there are opportunities for other people out there to do it. That market is still there, it is still being served; it is just not being served by the Disneys and Ubisofts and Activisions of the world.

JJ: *Is there also something being lost? Are there types of games that cannot be made any more?*

WS: I don't believe that. There are tools made available to everyone. There are lots of really good tools available to everybody who wants it, and there are so many ways to reach an audience now that the indie game movement is gaining traction. Suddenly you have Nintendo, Microsoft, and Sony all desperate for the broadest range of downloadable content they can get. Steam, Manifesto games, Shockwave, Yahoo...there are so many ways...Facebook. If you are a person with a dream and a game that you burn to make, the opportunities to make it and sell it are there. If you want to make the 100-hour game, you will find an audience for ten thousand people and go for it. Things like *Portal, flow,* the *Night*

Journey game that Tracy Fullerton did—those are not only reaching an audience, some of them are reaching a pretty big audience, and they are changing the mainstream industry. They are doing exactly what an indie movement should do, which is awesome. They are taking chances that I can't afford to take.

JJ: It is interesting what you say—that it is not that über-hardcore games are going away; it is more just that different people will be making and distributing them. You talked about making games for thirteen-year-olds who didn't have that sense of history and by history you meant video game history?

WS: Consider *BioShock* or look at Blizzard's entire oeuvre. There is nothing new in there, they execute exceptionally well on well-understood ideas. You can argue that BioWare does some of that. There is nothing wrong about it. God knows I am envious of the success of all of those companies. But for most players an immersive simulation is *BioShock*, for most players, *World of Warcraft* is MMOs. It is hard as a gamer to wrap your head around it, but most people who play *World of Warcraft* don't even know what *EverQuest* was. There is a training that has to go on. I could just assume that people who played *Ultima VII* had played the other six. And that they had played *Might and Magic*, and that they had played *Wizardry*. They understood the conventions of the genre and the medium, so there was all this shorthand stuff that I didn't have to worry about. Now, I can't make that assumption. *World of Warcraft* has to train people in what it means to be part of an MMO. *BioShock* had to train people in what it meant in a world where you really had some control over your actions, and that is a little bit different. This is something you have to pay a little more attention to now.

JJ: Isn't there something about the innovation-cloning balance? Developers want to be innovative. That seems to be at odds with the fact that people only know certain things and that people only have limited amounts of patience. Is there a balance there?

WS: My studio has a manifesto. One of the tenets is that innovation is a core value. I have no interest whatsoever in making a version of ideas that already exist. Every game I work on has to have at least one thing where I can say that no one on the planet has done that before. As long as we have that, I personally am happy with that. You are always finding balance—there are a variety of kinds of creativity. Someone like Will Wright is a blank-page designer. He opens his eyes one day and there is something completely new to the world that he has to make. I am kind of a synthesizer. I take well-understood elements and put them together and throw in my one new thing. And I am off to the races. In my own work there is clearly a balance between borrowing and innovation. I think everybody is kind of like that.

JJ: *The Wii [games] and* Guitar Hero *innovate by borrowing material from outside video games that players already know. Do you think that is how they can be innovative and accessible at the same time?*

WS: There are two interesting things in relation to that. One is to take things that people already know. I came to that revelation while working on *Deus Ex*. I realized at that point that I had spent most of my life and career trying to make people be interested in stuff that I thought was cool. And I realized how risky that was, how prone to failure that was, and how stupid it was. So now I go looking for what do people care about. That was the genesis of *Deus Ex*. There was this millennial conspiracy stuff. Everywhere I turned it was conspiracy this, conspiracy that, on TV, on the web, in books. I thought: people really care about this stuff. I will make a game about that. I think that is just smart. Finding things that people already care about and building on that foundation is great. And I completely agree that Harmonix and Nintendo nailed that. The other thing about both *Guitar Hero* and the Wii is that they force innovation through the peripheral. There is a truism in the game business that you cannot base your success on a peripheral that you can't guarantee that everyone will have. *Guitar Hero* and the Wii proved it wrong by packing the peripheral with the game.

JJ: *I sometimes hear the idea that in order to make games more accessible, you need more story, more things to identify with, and it actually seems that it is completely the reverse. That most of the games that are widely popular are very openly gamey and very openly artificial and have lots of user interface elements in your face and things saying "This is a game, you are doing well, this is a game." Why is it like that?*

WS: I think there are a lot of forces at work. For one thing, story requires commitment and implies attention over time. More gamey games don't require that commitment, which I think plays into the fact that people have more competition for their time and attention than they used to, so I think there is room for both. I think the gamey game that doesn't require commitment encourages the five-to-ten-minute play session and then you can put it down; I think that more closely matches life as it is lived in the twenty-first century. I want to check my e-mail, then I check Facebook, then I watch a few minutes of *Real World*, then I've got to go eat something, darn it, then I got to go to class. . . . That is not conducive to storytelling. But story is one of the things I was talking about earlier, about the things that people are already interested in. People have thousands of years of telling each other stories and they understand stories, stories are a good way of structuring experience and provide context and significance for experience. I think there is a place for story. Nobody is playing *Grand Theft Auto* for the story, but the fact that it is there plusses the experiences, makes it a bit more.

JJ: It is not that there is no story in most popular games, but it is story as a setting that gives some context, it is not structured around "then this happened, then this happened."

WS: I said this for years: "Games are not books, they are not movies, they are not plays." We have adopted storytelling conventions from other media. There must be a researcher somewhere to find a sixth way of telling stories in games. Maybe it is the pure contest, maybe it is *Guitar Hero*, letting you rise through the ranks in *Rock Star* and opening up bigger venues. Maybe that is games' dominant storytelling. Maybe my progress on my *Wii Fit* chart is the story. Maybe that is it, maybe Harmonix and some others already stumbled on the sixth story type and I have been thinking too much.

An Interview with Margaret Wallace

Margaret Wallace is a developer of downloadable casual games. She worked in the company PF Magic developing the games like *Dogz*, *Catz*, and *Babyz*, later cofounded Skunk Studios, and has recently moved on to cofound a new casual games studio in New York City called Rebel Monkey.

JJ: What do you think is the best starting point? Is it "Tell me about casual games" or "Tell me about casual players"?

MW: I think it is interesting to talk about the players. A lot of assumptions are made about who those players are, and a lot of people are trying to get into the head of players. People often spend a lot of time trying to define what a casual game is, but to me the most interesting things are who the players are, what their habits are around playing, and how casual games occupy a space as a leisure activity.

JJ: So what is a casual player?

MW: A casual player can be anyone. That is the easiest response. A hardcore player can also be a casual player. A casual player can be of any age group, any gender. They tend to enjoy playing games that are not punishing, tend to play games with themes that are a little more tied into mainstream culture, and tend to be less violent. A person who plays *Medal of Honor* can also be a casual player, it just depends on where they play. The person who plays a hardcore game might also be a casual player at work, if they are blowing off steam during lunch hour by playing a quick puzzle game.

JJ: Are you saying that it is not as much about having an identity as "a casual player," but it is more that sometimes you will be playing casually?

MW: Some people will be playing casually more often than others. What is really interesting about casual players is the fact that they don't self-identify as game players. Compare to when people go to the movies—they do not self-identify as "movie-goers." Like with movies, casual games occupy a space in their lives where casual players don't feel they have to identify their relation to them because these activities are so ubiquitous, such natural parts of their lives, like turning on the television and watching a television show. I believe many players of casual games also wouldn't identify themselves as being gamers because there may be a slight stigma associated with that. If you are talking about your average 40- or 50-year-old casual player, they might have associations of the word *gamer* as being a 14-year-old boy, or that it implies that you are slacking off and playing games all the time. I think it a cultural perception that we can turn to our advantage, by showing that casual games are not a niche experience, but part of everyday life, as normal as surfing the Internet, going to the movies, as watching television.

JJ: What is a casual game? Or do you have a problem with the term casual game?

MW: A lot of people feel that the phrase *casual games* does not do any justice to the kind of games we make. I can see those arguments, but I prefer not to get caught up in semantics. The term *casual games* has taken hold for now. I don't think that is the term that will be used forever. I think that maybe in a few years there won't be a need to distinguish a casual game from a hardcore game. When audiences become more sophisticated that may be deprecated. For casual games the most overused way of describing them is to say that "casual games are easy to learn, but difficult to master."

JJ: But that sounds like checkers.

MW: Right, but there is a shred of truth in referring to casual games like that because, across the board, casual games are inviting, the game play you need to master initially to get through a least the first few levels is made evident, the controls are as simplified as you can possibly get them, the duration from starting something to actually achieving your goal or being rewarded is much shorter. Casual games reward players much more often, and casual games are much more forgiving than other kinds of games. It is always a balance between making a game that is too easy and making a game that is challenging. I have definitely made games that were too easy and didn't do well because ultimately the audience did not feel compelled to keep playing. I have also been involved in designing games that were too difficult for the audience, and they also did not do too well. It is a huge design challenge. You are working with such limited variables compared to a game that would take twenty or thirty hours to play and it really constrains you in terms of experience and the mechanics you create and communicate to the player.

JJ: Where do you see the beginning of casual games?

MW: If we are talking about digital games as opposed to board or card games? You could say that parlor games were the first casual games, tag, etc. In terms of digital games, a lot of people point to the early arcade games and Atari games. I think that only holds water as far as the duration of the experiences were limited, and in terms of what you could actually do. But...I would not call those games casual games because the early arcade games were probably too hard. I would always go back to *Tetris*. That is more like the first casual game. I do think the virtual pet projects I worked on were also among the earliest casual games, or activities. But to me, *Tetris* is the first breakout successful casual game. But if a person made *Tetris* today, it might ironically not do so well in the casual game market compared to other titles.

JJ: Why not?

MW: I think to the audience targeted by casual games in the United States today, *Tetris* would be boring, or not immersive enough. I know that people have tried to do *Tetris*-style games in the downloadable casual space recently, and they have not performed well. *Tetris* continues to perform well on mobile phones though.

JJ: When you make games, do you have an image of the audience in your head? What is the level of detail? The TV shows they like? The way they shop?

MW: I feel fortunate that while being in the casual game portion of the game industry the last seven to eight years, I have had direct access to my user base. When you have your own website, you hear the opinions from your customers. I have also had the opportunity to do survey research on my users. I get inspired by things happening in popular culture, but I will always try to hone my ideas with my audience in mind. Will this fly in Peoria? Will it work for people from the Midwest? Will people from the Midwest be offended by what I make? Somebody from Ubisoft once said to me that casual games were the most vanilla-themed content they had ever seen, and they had never in a million years considered they would be popular. I think a lot of the people who play these games are playing at work or they are playing with their children, or grandchildren, or it is a family thing, and it is the equivalent of prime time television. You would not put risqué content on TV at 8 o'clock at night, and you don't do that with casual games.

I also look at the research that is being done, like the research Real Networks does which matched very well with my own informal research. I will assume that casual players may like Oprah, some of them might like Tyra Banks. I try to see what cultural forms are prevalent, and what might resonate for what I consider to be the typical casual game players.

JJ: Do you think there were some lessons about audiences that board game developers like Parker Brothers knew, but that video game developers forgot over time?

MW: Definitely. When the company I was working at, PF Magic, was bought by Mindscape, and we were integrated into a traditional console gaming company, including a company called SSI that made games like *Panzer General* and all the war games, and Brøderbund which made *Prince of Persia*, and we had *Warhammer* and those titles. People at that company did not get us, they did not like us, they thought we were a joke. The traditional games industry did not know how to handle our games, *Dogz* and *Catz*, virtual pets. The sales people had to place our games in retail along with *Panzer General*. They did not know what to do with us. The marketing people did not know how to handle us. Something had happened in the game industry when I joined it, or just before, where people had discovered graphics. A lot of the original tenets that made something like Parker Brothers got lost. Cool graphics were used and still are to mask a lot of bad game design and bad ideas. A lot of the elitism in the game industry about who to target comes from that. When I was at PF Magic, we were making *Dogz* and *Catz*, we would get data from our customers saying that 14-year-old boys were not the dominant users of the games, it was girls and women. It was like a paradigm shift that people had the hardest time getting over. Even at PF Magic when we looked at the data, and saw that our users were more balanced in terms of the gender breakdown, people had the hardest time accepting that. We finally started talking about it and then we got acquired by Mindscape and then everything went bad. It was like an elitist club. People did not recognize that there was a wider player base out there.

JJ: Where do you see the balance between innovation and cloning?

MW: I always laugh at the cloning question, because casual games are small-sized games, so it is easy to pick out what is cloning, what is derivative. But if I look at popular culture, mainstream culture, I see nothing but clones. When I see *EverQuest* and *World of Warcraft*, I see cloning going on. But it is easier to see with casual games because they are smaller titles. The boundary between cloning and innovating is something that everyone [who] makes video games faces. If it is a thirty-million-dollar game, a one hundred thousand- or a three hundred thousand-dollar game. The balance is that you are never going to make an AAA blockbuster hit if you are just cloning. You may pick up some scraps, but you are not going to be a huge success just cloning. You may be a tier-two developer this way and make a living, but you are never going to propel yourself into the stratosphere. You cannot ignore the game mechanics that resonate with people, and if you are working with a mouse button there is not a lot you can really do, but to make blockbuster hits or to make AAA hits, there has to be something about what you are doing that innovates. Whether it is in the mechanics or the narrative or how they are integrated.

The portals are in an interesting situation: although they hold the power and the value chain, and they have the access to the audience, they are clearly relying on the content creators to help them retain the audience and differentiate their channel from other channels. There is a vicious cycle where the people with access to the audience are making demands on the kinds of content being made, not wanting to take a lot of risk due to the quarterly numbers, but at the same time developers don't want to take risks because they want to appear on the portals. That is why I think innovation will not come from these relationships; innovation will always come from something somebody tried on the fringes.

JJ: One of the things right now seems to be genre churn, with the dominant genres changing. Is that the way it will go on forever?

MW: I think there will always be breakthrough genres, so that will go on forever. I think wider audiences will go to different genres. But right now it is the I spy/hidden object games that are dominant, and I think that will go on for a while.

JJ: *I find it interesting that the I spy/hidden object games are lifted completely from a nondigital genre. Are there a large amount of nondigital games that are simply waiting to become casual games?*

MW: I think so. I have seen nondigital games being made into casual games and performing poorly. So it is really about finding the right games to make into digital games. There are no guarantees; it is all in the implementation.

JJ: *What do you think about the I spy games? I have enjoyed all casual game genres, but this, I simply cannot tolerate. What do you feel about these games?*

MW: Well, we are not making an I spy game though we did have an idea for one that would have been different. I get sucked into those I spy games and have suddenly played them for an hour. I check them all out because I play the current top games. They do draw me in, but it is probably because they are relaxing. Something about them makes me relax. That is the way my brain works when I play casual games. I get great satisfaction from looking around and hunting for hidden objects.

JJ: *Just now I realize that the reason I don't like these games is that I never feel clever in those games, I always feel stupid. Do you feel clever playing these games?*

MW: Yeah, I do. I feel clever in those games. I enjoy finding hidden things.

An Interview with Eric Zimmerman

Eric is a game designer who has worked in the game industry for more than fifteen years. He was cofounder and chief design officer of the game studio Gamelab, developers of *Diner Dash* and other titles. Eric is coauthor with Katie Salen of *Rules of Play* (MIT Press 2004).

JJ: *The first question I have on paper is, "What is a casual game?" but I know you don't like that term. Why don't you like it? What is the issue?*

EZ: Because as a producer of culture, I like to think that my audience can have a sort of deep and dedicated and meaningful relationship with the works that I produce. And the notion of a casual game implies a kind of light and less meaningful relationship to the work. That's not saying that so-called casual players can't be quite dedicated, but what the term points to is what I feel are negative connotations. That's why I don't like the term *casual game*. If I were a musician, I would never be proud of someone saying, "I make casual music." In the same way, I don't feel that it's complimentary to say that one makes casual games. On the other hand, I acknowledge that it is a term in common parlance, and it's a useful term. So, I'm not saying it should be abolished; I'm just registering my disdain for it. In terms of defining it, it can be defined in many ways. Typically, how it's used is . . . to refer to certain types of design, to refer to certain types of platforms and revenue models, to refer to certain audiences. I think those are the main ways, audience, design, and then sort of business model/platform.

A lot of people use casual games to mean online, free web games or downloadable games to exclude console platforms; other people say, "Well, the Wii is a casual platform"; still other people say, "Well, on certain consoles . . . *Guitar Hero* is a casual game." But then, they probably mean the audience. I'm not interested in proposing a single term to combat that. My main thing is that I don't want people thinking of what I do as casual. It's a pejorative word—what musician would like to say, "I make casual music"?

JJ: I have heard you use the term mass-market games *instead.*

EZ: Right, and by mass market, with the word *market* in it, I'm trying to make it clear that I am talking about audience.

JJ: Do you see reorientation happening in the game industry, games reaching a larger audience than they did previously?

EZ: I think that the audience for game players is expanding on several different fronts. On the youth front—as opposed to when you and I, who are now on our late thirties, were kids and video games were popular but also somewhat of a geeky pastime—a study just pointed out that video games are played by 97 percent of kids, and so it is a part of their media landscape in a way that is as pervasive as television. I just saw a panel of kids at a conference, one of the questions that they were asked was "If you had to give up either television or video games, what would it be?" Now, these were not hardcore players. They were sort of a mix and there were about a dozen of them, and only one of them said they would give up video games. The rest of them all said they would more easily give up television than video games. So, that is one front in which the audience is expanding. As new generations come up, the idea of playing games is not something that is going to be seen as particularly geeky. I was amazed by how these teenagers aged about 13 to 17 already had multiple layers of nostalgia and connoisseurship of earlier games. In a large part, because their budgets are very limited and so they buy a lot of old games. One girl who was 14 said, "Oh, I'm really not a game player. I'm really very casual. It's my brother that plays the games." She actually used the word "casual." But then, when she was pressed about what game she actually played, she said, "Oh well, you know, it depends, certain moods I'm in, I like to play certain games. If I'm with my friends, we'll get competitive; if I'm alone and I just want to mellow out, I'll get my old Game Boy Color out and play some *Super Mario* for an hour or two". There's nothing casual about that. Of course, there's the so-called traditional casual audience which are women in their thirties and forties [who] were not traditional gamers. Then there are so-called lapsed gamers or people who grew up playing games but no longer have the amount of time to put into playing serious hardcore games. This may be more of a male audience, more of a twentysomething, thirtysomething audience. Once you leave college, your time starts evaporating for those sorts of pursuits. There are many different audiences and that is exciting to me. I mean, I think the idea that the audience is broadening for casual games is connected to broader movements in industrialized society and culture that have to do with our culture becoming more ludic, more oriented toward playing. As digital technologies and networks of information, the Internet, computers, mobile technologies, more and more pervade our lives, [and] the ways in which we socialize and flirt and communicate and learn and work and do our taxes and engage with our government and manage our finances, and many, many other important aspects of our lives, the more I think our culture becomes primed for play and particularly, games as the dominant form of leisure. Because games are the form of culture that is most intrinsically related to those things, to systems, technology, information, and mediated communication. It wouldn't surprise me [if], just like society in the twentieth century gave rise to cinema...and television, in this newer century where information technology is now being supplanted by ludic technology,...play becomes a more dominant paradigm for culture rather than the moving image.

JJ: I am working from the assumption that there is a basic change in the video game industry, that games are reaching a broader audience. What do you think is the best starting point for discussing that? Is it to start with casual games or is to start with casual players?

EZ: I would say the starting point for thinking about that question is who is playing games. If you look at the long history of games, the millennia-old history of games, we see that games and play are a near universal in human culture. In a culture like ours, not everybody plays games their whole life but if you look at every game-playing activity, poker, sports, organized sports, informal sports, leisure activities like skiing, crossword puzzles, *Scrabble*, board games, playing games among kids or with kids, educational games played in the classroom and electronic games, there's such a wide variety of game activities in industrialized Western culture. Obviously, every culture has their own universe of games that are played for different reasons and play activities which might be slightly less formal forms of games. Technology has influenced games like the printing press influenced board games that could [then] be published in full color. With digital games, you have to look at the culture out of which they came. Computer culture really had very geeky origins, which I think it has taken decades to shake off. My personal anecdote about this is that around 1995 I was living in New York City, working in my first game company and I read a personals ad. This was already a few years after *Wired Magazine* came out, started to make technology sexy, this was during the CD-ROM heyday in New York City. Companies like Voyager were making coffee table CD-ROMs and geek chic was becoming chic, actually. So, I remember reading a personal ad that said, "I'm looking for a man of this age, this and that. Single, white, female." at the end it said, "No computer nerds, please." And I had this weird sense of vertigo inside. And I realized that for me computers and technology were obviously sexy but for this person, there was just such a stench of unattractiveness to computer technology. That was such a given that they didn't want to consider dating someone like that. What was interesting though was that even by the mid-1990s computers video games had been through a bit of a cultural renaissance. I was born in '69 so during the eighties I was in junior high school and high school. At that time *Pac-Man* was popular: there was a television show about *Pac-Man*, there was a hit song about *Pac-Man*. Video games were enjoyed by lots of people, were accepted by lots of people, and if anything, they were seen as transgressive, which is a little bit the opposite of nerdiness. Arcades had replaced the pool halls of the fifties and sixties and teenage delinquents who were skipping class went to hang out at video arcades. But it is a complicated question. I don't think that it is a single trajectory where they were unpopular and geeky and now they are becoming more popular. I think it's actually that they have gone through several stages in transformations.

JJ: You have expressed some dissatisfaction with the downloadable casual games channel.

EZ: I gave a talk titled "Casual Games Are Dead." By that I mean that they are dead not for the publishers, distributors and aggregators, but for the game creators and developers. Dead financially and creatively. Financially because even though the audience has grown, conversion rates have gone down. Churn on the portals is way up and so there are fewer successes, and big successes are much smaller than they were several years ago. *Diner Dash* was on the top ten list for more than a year, whereas now a triple A hit game will be on the top ten lists for months or even weeks at most. It is just harder to make money. Budgets have gone up overall. There also are more developers competing for publisher dollars, and some of them work in cheaper countries than Western Europe and the United States. I'm thinking of Asia, Eastern Europe, and South America. Gamelab then is competing against companies that are trying to do the same thing more cheaply. Overall, it's very hard commercially, the portals are very mercenary and they don't really publicize their back catalog unlike someone like Amazon where they're trying to use collaborative filtering and trying to get a long tail. With downloadable casual games it is all about whatever is the latest game that is converting well. When that doesn't happen any more, it goes to archives and is essentially lost. Those are the commercial reasons.

JJ: *What are the creative reasons?*

EZ: When downloadable casual games started, the optimism was that great things would come of making small-scale budget games rather than making big, multimillion-dollar hardcore games. Digital distribution, not paying for shipping and retail shelf space would lead to a meritocracy also due to the "try before you buy" model. There was an idea that downloadable games could be a renaissance for innovation in terms of theme, content, and gameplay. But in fact, the downloadable casual games industry has evolved into something *more* clone-driven and genrebound than the so-called hardcore game industry that it sought to make an end run around. The downloadable casual games industry has become a parody of itself. I think that there are little pockets of web games and independent games that are interesting, but nobody is making money from those.

It is hard to innovate in that realm. If that is what players want that is fine. Just like daily newspapers, tabloid papers dominate over so-called more serious journalism. If players want to pay for more hidden object games or *Diner Dash* ripoffs, then that's what they want to play and that's fine. I'm not complaining about the tastes of players, I'm just saying if I were a journalist I would rather work at the *New York Times* than at the *New York Daily News,* because I'm pretentious that way. As a game designer, I don't want to make clones of existing games. I would rather work in a more, let's say, serious place where I feel I can do more innovative work and works with more lasting value. That's part of why Gamelab isn't really doing any more downloadable games.

JJ: A few years ago you were telling a story about a slightly surreal game that the game portals would not publish.

EZ: One of the major portals said, "We cannot post a game that might potentially offend any member of our audience." Imagine a record company saying that. There would be many, many genres of music that would have to be removed, whether it was country singers singing about conservative politics, or whether it was rap artists rapping about life in an urban environment.

Notes

1 A Casual Revolution

1. Namco, *Pac-Man*, 1980.

2. Blizzard Entertainment, *WarCraft III*, 2002.

3. Konami, *Dance Dance Revolution*, 1998.

4. Harmonix Music Systems, Inc., *Guitar Hero*, 2005.

5. Harmonix Music Systems, Inc., *Rock Band*, 2007.

6. Sandlot Games, *Cake Mania 3*, 2008.

7. Preece, *Desktop Tower Defense*, 2007.

8. Ryan, "Game Designers Focus on Girls," 2008.

9. Interview with Warren Spector, see appendix C.

10. Entertainment Software Association, *Essential Facts*, 2008.

11. Pratchett, *Games in the UK*, 2005.

12. Boyer, "NPD: 72% of U.S. Plays Games," 2008.

13. Lenhart et al., "Teens, Video Games, and Civics," 2008.

14. Gartenberg, *Simple Games*, 2007.

15. Dobson, "Study: 'Casual' Players Exhibit Heavy Game Usage," 2006.

16. See the player survey in appendix A and the developer interviews in appendix C.

17. There is a growing body of research on games and gender, such as Cassell and Jenkins, eds., *From Barbie to Mortal Kombat*, 1998; Y. B. Kafai et al., *Beyond Barbie*

and Mortal Kombat, 2008; Enevold and Hagström, "My Momma Shoots Better than You!," 2008; Nakamura and Wirman, "Girlish Counter-Playing Tactics," 2005; Flanagan, "Troubling 'Games for Girls,'" 2005; Fron et al., "The Hegemony of Play," 2007. This research has dealt almost exclusively with the question of girls and women in relation to games, the exceptions being Jenkins, "Complete Freedom of Movement," 1998 and Burrill, *Die Tryin'*, 2008.

18. Industry veteran Steve Meretzky has argued that the video game industry began losing contact with the casual audience around 1980, then reintroduced a separate category of games for a casual audience with the inclusion of Solitaire in Windows 3.0 in 1990, and that hardcore and casual games are now merging again (Meretzky, "What Is a Casual Game?," 2007).

19. Interview with a player of downloadable casual games, see appendix B.

20. Addison, "Confessions of a Non-Gaming Mom," 2008.

21. Howson, "Casual Games Rule Charts," 2008.

22. Microsoft, "Microsoft Reveals First Details of Next-Generation Xbox," 2005.

23. Sony Computer Entertainment Inc., "SONY COMPUTER ENTERTAINMENT INC. TO LAUNCH ITS NEXT GENERATION," 2005.

24. Gibson, "Phil Harrison on the Future of PlayStation Interview," 2005.

25. Costikyan, "Burning Down the House," 2005.

26. Since consoles have very different technical designs, there is no simple way to compare their computational or graphical power. I have here compared the raw clock frequency of the console CPUs. I have ignored that the Xbox 360 and PlayStation 3 have CPUs with multiple cores, and that CPUs are not equally effective per clock cycle.

27. VGChartz.com. "VGChartz.com," 2009.

28. According to a recent survey, 50 percent of consumers report price to be an important factor when purchasing a console, but only 11 percent reported high-definition graphics to be important (eMarketer, *Importance of Select Factors to US Console Video Gamers*, 2008). In the not-too-distant future, all video game consoles will undoubtedly feature high-definition graphics, but as of now high-definition is the prime example of how a technical selling point has failed to excite a broad audience.

29. In addition to the Nintendo Wii leading the current console generation, Nintendo's DS handheld game console is also outselling the Sony PSP handheld console—even though the PSP has technically better graphics.

30. Atari, *Pong*, 1972.

31. Psygnosis, *Wipeout*, 1995.

32. PopCap Games, *Bejeweled 2 Deluxe*, 2004.

33. Nintendo, *Wii Sports*, 2006.

34. Salen and Zimmerman, *Rules of Play*, 2004.

35. Taylor, *Play Between Worlds*, 2006.

36. Sony Online Entertainment, *EverQuest*, 1999.

37. Consalvo, *Cheating*, 2007.

2 What Is Casual?

1. Oberon Media, "GAME DEVELOPERS SUBMISSION GUIDELINES," 2007.

2. A discussion on the industry "IGDA Casual Games SIG Mailing List" in early 2007 yielded inconclusive results about the origin of the term "casual games."

3. Kim, "Games for the Rest of Us," 1998.

4. Personal correspondence.

5. Kim, "Games for the Rest of Us," 1998.

6. The WizardWorks Group, Inc., *Deer Hunter*, 1997.

7. Lohr, "Computer Games Venture," 1999.

8. Interview with Eric Zimmerman, see appendix C.

9. Finnish game researchers have pointed to six different uses of *casual* to name casual game culture, casual games, casual gaming, casual playing, casual gamers, and casual game players (Jussi Kuittinen et al., "Casual Games Discussion," 2007). For example, casual can be seen both as an identity (being a casual player) and as a specific way of playing (playing casually).

10. Range, "The Space Age Pinball Machine," 1974.

11. Los Angeles Times, "New Video Games Sweep Nation," 1974.

12. Harmetz, "Is Electronic-Games Boom Hurting the Movies?," 1981.

13. Midway, *Ms. Pac-Man*, 1981.

14. Goldstein, "Why Is Pac-Man Grinning?," 1982.

15. Pajitnov and Gerasimov, *Tetris*, 1985.

16. In the interview in appendix C, Margaret Wallace argues that *Tetris* can be considered the first casual game.

17. Meretzky, "What Is a Casual Game?," 2007.

18. Cyan, *Myst*, 1993.

19. Pearce, "The Truth about Baby Boomer Gamers," 2008.

20. McElroy, "Behind the Game: Bejeweled®," 2006.

21. I interpret *core* gamer as being a synonym of *hardcore* gamer here.

22. USA Today, "Nintendo Hopes Wii Spells Wiinner," 2006.

23. Marriott, "The Un-Doom Boom," 2003.

24. Schell, "Hardcore Games for Casual Audiences," 2008.

25. Kapalka, "10 Ways to Make a BAD Casual Game," 2006.

26. Švelch, "YOU BURNOUTS ROTTING IN FRONT OF YOUR COMPUTERS!," 2008.

27. Kotaku, "Kotaku, the Gamer's Guide," 2007.

28. Dobson, "Study: 'Casual' Players Exhibit Heavy Game Usage," 2006.

29. Player interview, see appendix B.

30. Epic Games, *Gears of War*, 2006.

31. Lane, Chua, and Dolan, "Common Effects," 1999.

32. Nielsen, *Usability Engineering*, 1993.

33. We have no data on the extent to which casual game developers are versed in interface design literature.

34. Beaudouin-Lafon, "Instrumental Interaction," 2000, 448.

35. Scott, *Jenga*, 1986.

36. Electronic Arts Los Angeles, *Boom Blox*, 2008.

37. Juul, Norton, "Easy to Use and Incredibly Difficult: On the Mythical Border between Interface and Gameplay," 2009.

38. Juul, *Half-Real*, 2005.

39. Shneiderman, "Direct Manipulation," 1983.

40. HipSoft, *Build-a-Lot*, 2007.

41. Goble and Price, "Build-A-Lot Post Mortem," 2008.

42. Nielsen Games, *Video Games in Europe 2008*.

43. Rockstar Games North, *Grand Theft Auto: San Andreas*, 2005.

44. PopCap Games, "Survey: Tens of Millions of White Collar Workers Play 'Casual' Video Games," 2007.

45. Kapalka, "10 Ways to Make a BAD Casual Game," 2006.

46. Gamelab, *Shopmania*, 2006.

47. Herdlick and Zimmerman, "Redesigning Shopmania," 2006.

48. iWin, *Jewel Quest II*, 2007.

49. Bell, "Jewel Quest 2 Review," 2008.

50. Suzie Q., "Jewel Quest 2 Reviews," 2007.

51. E-mail interview, see appendix B.

52. E-mail interview, see appendix B.

53. Interview with Margaret Wallace, appendix C.

54. Juul, *Half-Real*, chap. 3.

55. Juul, "Fear of Failing?," 2008.

56. Game designer Sheri Graner Ray claims that the question of punishment is *inherently* tied to gender: "The female style is the forgiveness for error with no irretrievable loss, whereas the male model is punishment for error." Kafai et al., *Beyond Barbie and Mortal Kombat*, 2008, 322. I find this claim highly dubious for a variety of reasons, the least of which is that many female players also enjoy highly punishing games, as documented in appendix B.

57. Smith, *Manic Miner*, 1983.

58. Friends Games, *Magic Match*, 2007.

59. Gamelab, *Diner Dash*, 2003.

60. PopCap Games, *Peggle*, 2007.

61. Norman, *Emotional Design*, 2005.

62. Norman, *The Design of Everyday Things*, 2002.

63. Norman, see note 61, 65–69.

64. Gabler et al., "How to Prototype a Game in Under 7 Days," 2005.

65. Epic Games, *Gears of War*, 2006.

66. Note that there may be many games promoted as casual that do not match this list, and there may be games that match this list that are not promoted as casual games.

67. Dobson, "Study: 'Casual' Players Exhibit Heavy Game Usage," 2006.

68. Compare this to Jason Mittell's discussion of how it is unclear to whom the word *audience* refers (Mittell, *Genre and Television*, 2004, 94–95). Who are the casual players? Is it those who spend the most time playing casual games? Those who spend the *least* time playing casual games? Those who spend the most time on casual game websites?

69. Malaby, "Beyond Play," 2007, 95.

70. Aarseth, "Playing Research," 2003, 7.

71. Mikael Jakobsson's study of *Super Smash Bros Melee* is a good example of a player-centric view of games. He observed how players made their own house rules in the game. Based on this, Jakobsson concludes, "As we have seen in the example of [*Super*] Smash [*Bros Melee*], the very nature of a game can change without changing the core rules." In Jakobsson, "Playing with the Rules," 2007.

72. The game-centric view would assert that players or contexts have no influence on how a game is played or experienced. This position may not actually exist, though economic game theory (Neumann and Morgenstern, *Theory of Games and Economic Behavior*, 1944) is occasionally described as such. More inclusive game-centric views can be found in game design texts, obviously focused on the development of a game rather than on players or contexts. Even so, the role of the game designer is sometimes described as being an "advocate for the player" (Fullerton, Swain, and Hoffman, *Game Design Workshop*, 2004, 2). Theoretical game studies have produced game-oriented scholarship on such topics as game structures (Juul, "The Open and the Closed," 2002) or game typologies (Aarseth, Smedstad, and Sunnanå, "A Multidimensional Typology of Games," 2003).

73. Q Entertainment, *Lumines Live!*, 2006.

74. Q Entertainment, *Lumines*, 2004.

75. Interview with Margaret Wallace, see appendix C.

76. Ibid.

77. Ibid.

78. Nintendo, *Financial Results Briefing*, 2008.

79. Casamassina, "Wii Sports Review," 2006.

80. Blizzard Entertainment, *World of Warcraft*, 2004.

81. Ducheneaut et al., "Building an MMO with Mass Appeal," 2006, 289.

82. Lafferty, "World of Warcraft Review—PC," 2004.

83. Nintendo EAD Tokyo, *Super Mario Galaxy*, 2007.

84. Olsen, "Anti-Peggleite," 2008.

85. Player interview, see appendix B.

3   All the Games You Played Before

1. Infinite Interactive, *Puzzle Quest*, 2007.

2. Hatfield, "Puzzle Quest," 2007.

3. Steinberg, "Puzzle Quest," 2007.

4. sinofsky, "If You Love RPGs and Puzzle Games," 2007.

5. Carless, "In-Depth," 2008.

6. ELIZABETH "LIZZIE." "FUN? WELL.... DIFFERENT.... THAT'S FOR SURE !!," 2008.

7. Mittell, *Genre and Television*, 2004, chap. 1.

8. Big Splash Games, *Chocolatier*, 2007.

9. Big Fish Studios, *Azada*, 2007.

10. For more detailed discussions of mechanics, see Järvinen, "Games without Frontiers," 2008, and Sicart, "Defining Game Mechanics," 2008.

11. Costikyan, "Game Styles, Innovation, and New Audiences," 2005.

12. The "simulation" genre is the counterexample to Costikyan's claim that genres are only defined by mechanics. For example, the review site GameSpot lists train simulators, flight simulators, and high school simulators in the simulation category.

13. Juul, *Half-Real*, 2005, chap. 2.

14. Parlett, *A History of Card Games*, 1990, 157–161.

15. Cadogan, *Illustrated Games of Patience*, 1876.

16. Jones, *Games of Patience for One or More Players*, 1898, 1.

17. A Solitaire game on a computer is an *implementation* (rather than an adaptation) of nondigital Solitaire because every possible action and game state within the card game has a corresponding action and game state in the computer-based version. Juul, *Half-Real*, 2005, chap. 2.

18. Tarbart, *Games of Patience*, 1901, preface.

19. Cadogan, *Lady Cadogan's Illustrated Games of Solitaire or Patience*, 1914.

20. Microsoft, *Solitaire*, 1990.

21. Kallio, Kaipainen, and Mäyrä, *Gaming Nation?*, 2007, 79.

22. Meretzky, "What Is a Casual Game?," 2007.

23. Parker Brothers, *Monopoly*, 1936.

24. Orbanes, *The Game Makers*, 2003, 13.

25. Cerny and John, "Game Development," 2002.

26. Ibid., 36.

27. Davis, Steury, and Pagulayan, "A Survey Method for Assessing Perceptions of a Game," 2005.

28. Clark, "Core Values of the Casual Industry, 2006, 19.

29. Interview with Garrett Link, see appendix C.

4 Innovations and Clones: The Gradual Evolution of Downloadable Casual Games

1. Telephone interview, see appendix B.

2. Interview with David Rohrl, see appendix C.

3. Relentless Software Ltd., *Buzz!*, 2007.

4. Big Fish Studios, *Mystery Case Files*, 2005.

5. PopCap Games, "PopCap's Bejeweled(R) Franchise Hits 25 Million Units Sold Mark," 2008.

6. PopCap Games, *Bejeweled 2 Deluxe*, 2004.

7. Dobson, "Study," 2006.

8. PF Magic, *Dogz: Your Computer Pet*, 1995.

9. PF Magic, *Catz: Your Computer Petz*, 1996.

10. Interview with Margaret Wallace, see appendix C.

11. Kapalka, "10 Ways to Make a BAD Casual Game," 2006.

12. Varney, "Attack of the Parasites," 2006.

13. PopCap Games, "Publishing FAQ," 2006.

14. Tunnell, "Five Foundational Steps," 2006.

15. Eliot, "Philip Massinger," 1920.

16. Culin, "Mancala," 1971.

17. Geryk, "A History of Real-Time Strategy Games," 2001.

18. Moribe, *Chain Shot!*, 1985.

19. Pajitnov and Gerasimov, *Tetris*, 1985.

20. McElroy, "Behind the Game," 2006.

21. Gamehouse, *Collapse*, 1998.

22. Intelligent Systems, *Panel de Pon*, 1995.

23. Nintendo, *Dr. Mario*, 1990.

24. Bulletproof Software, *Yoshi's Cookie*, 1992.

25. Intelligent Systems, *Panel de Pon*, 1995.

26. Taito, *Plotting*, 1989.

27. Mumbo Jumbo, *Luxor*, 2005.

28. PopCap Games, *Zuma*, 2004.

29. Raptisoft, *Chuzzle Deluxe*, 2005.

30. iWin, *Jewel Quest*, 2004.

31. IGDA Casual Games SIG, "2005 Casual Games White Paper," 2005, 55.

32. The amount of moderate innovation in the tree is to be expected given the tree is an attempt to find similarities between games.

33. Reflexive Entertainment, *Big Kahuna Reef*, 2004.

34. Hot Lava Games, *7 Wonders of the Ancient World*, 2006.

35. PopCap Games, *Zuma*, 2004.

36. For example, Phil Steinmeyer, "Playing It Safe (Enough)," 2005.

37. Mumbo Jumbo, *Luxor*, 2005.

38. Wildfire Studios, *Tumblebugs*, 2005.

39. Big Fish Studios, *Atlantis*, 2005.

40. Cifaldi, "Popping in on PopCap," 2005.

41. Mitchell, *Puzz Loop*, 1998.

42. Seydoux, "shokkingu hitofude," 2006.

43. Kuchera, "WiiWare and VC Releases," 2008.

44. Dillon, "Casuality," 2006.

45. Atari, *Centipede*, 1980.

46. Taito, *Puzzle Bobble*, 1994.

47. Barr, *Cubism and Abstract Art*, 1966.

48. Tufte, *Beautiful Evidence*, 2006, 65.

49. Mirabella III, "IGN," 2001.

50. MobyGames, "Tile-Matching Puzzle Games (Creation)," 2008; MobyGames, "Puzz Loop Variants," 2008; MobyGames, "Falling Block Puzzles," 2008.

51. Big Fish Games, "Puzzle Games|Big Fish Games," 2008.

52. Data was gathered through the Games Sales Charts service (J. C. Smith, "CasualCharts.com," 2008) using the "World map" function to list most popular mechanics based on the top-ten lists of popular portals. I treated the "match3" and the "chain popper" mechanics as matching tile games.

53. Gamelab, *Diner Dash*, 2003.

54. Big Fish Studios, *Mystery Case Files*, 2005.

55. Interview with Eric Zimmerman, see appendix C.

5 Return to Player Space: The Success of Mimetic Interface Games

1. Interview with a player, see appendix B.

2. Nintendo EAD, *Wii Fit*, 2008.

3. Exidy, *Destruction Derby*, 1975.

4. Midway, *Sea Wolf*, 1977.

5. Addison, "Confessions of a Non-Gaming Mom," 2008.

6. Technically, the 1989 Nintendo Power Glove for the Nintendo Entertainment System was also a generalized input device that sensed the motions of a player, but it was ultimately unsuccessful, with only a few games exploiting its capabilities.

7. Sumo Digital, *Virtua Tennis 3*, 2007.

8. Beaudouin-Lafon, "Instrumental Interaction," 2000, 448.

9. Neversoft, *Guitar Hero III*, 2007.

10. Interview with David Amor, see appendix C.

11. Totilo and Elias, "Slash, Developers Riff on 'Guitar Hero III,'" 2007.

12. Zelfden, "Gingold Talks Spore's 'Magic Crayon' Approach," 2008.

13. Flaherty, "Flaherty: Stay Cool," 2008.

14. Juul, "A Certain Level of Abstraction," 2007.

15. Namco, *Donkey Konga*, 2003.

16. Sonic Team, *Samba de Amigo*, 1999.

17. See appendix B.

18. Interview with Sean Baptiste, see appendix C.

19. Lazzaro, "Why We Play Games," 2004, 35.

20. Statistics collected on October 19, 2008.

21. Young, "The Disappearance and Reappearance and Disappearance of the Player in Videogame Advertising," 2007.

22. Nintendo EAD, *Wii Play*, 2006.

23. Ubisoft Montpellier, *Rayman Raving Rabbids*, 2006.

24. Skill is acquired faster in *Guitar Hero* in comparison to playing actual guitar because it is easier for the non-guitar-playing player to learn a given song *in the game* than to learn it on an actual guitar. Just as video games have learning curves, so do musical instruments.

6 Social Meaning and Social Goals

1. Interview with David Amor, see appendix C.

2. Garfield, "Metagames," 2000.

3. Nintendo EAD, *Animal Crossing*, 2002.

4. Onesound, *Animal Crossing Is Tragic*, 2005.

5. Järvinen, "Communities of Nurturing," 2007.

6. Smith, "Plans and Purposes," 2006, 217–227.

7. Personal conversation.

7 Casual Play in a Hardcore Game

1. Itzkoff, "Rec-Room Wizard," 2008.

2. Maxis, *The Sims 2*, 2004.

3. Rockstar Games North, *Grand Theft Auto*, 2005.

4. Apperley et al., "Researching Kids and Videogames," 2008.

5. Piccione, "In Search of the Meaning of Senet," 1980.

6. Juul, *Half-Real*, 2005, chap. 2.

7. Maxis, *SimCity*, 1989.

8. Juul, *Half-Real*, 2005, chap. 2.

9. Konami, *Scramble*, 1981.

10. Konami, "Scramble Installation Manual," 1981.

11. This reflects the game convention that goals are harder to reach than non-goals (Juul, *Half-Real*, 2005, 40).

12. Rockstar Games North, *Grand Theft Auto*, 2005.

13. Maxis, *The Sims 2*, 2004.

14. Maxis, *SimCity*, 1989.

15. DMA Design, *Grand Theft Auto III*, 2001.

16. Braben and Bell, *Elite*, 1985.

17. Microprose Software, *Pirates!*, 1987.

18. Nintendo EAD, *Super Mario 64*, 1996.

19. Mary Flanagan has interviewed *Grand Theft Auto* players who prefer simply driving around to completing missions (Flanagan, "Troubling 'Games for Girls,'" 2005).

20. Eco, *The Search for the Perfect Language*, 1995, 21.

21. Harmonix Music Systems, Inc., *Rock Band 2*, 2008.

22. Interview with Sean Baptiste of Harmonix, see appendix C.

8 Players, Developers, and the Future of Video Games

1. Interview with a *Guitar Hero* player, see appendix B.

2. Staiger concludes that "context is more significant than textual features in explaining interpretative events" (Staiger, *Perverse Spectators*, 2000, 3).

3. Interview with Jacques Exertier, see appendix C.

4. Interview with a player of downloadable casual games, see appendix B.

5. GameDaily Staff, "Are Big Budget Console Games Sustainable?," 2006.

6. Valve, *Half-Life 2*, 2006.

7. Valve, "Half-Life 2: Episode One Stats," 2008.

8. Preece, *Desktop Tower Defense*, 2007.

9. For players with only limited administration rights over their computer, such as players at work or at educational institutions, browser-based games may be the only ones to which they have access.

10. area/code, *Parking Wars*, 2008.

11. Singline and Hill, "Your Turn," 2008.

12. Eric Zimmerman says the following about content in the downloadable casual games channel (interview in appendix C): "One of the major portals ... said that 'We cannot post a game that might potentially offend any member of our audience.' Imagine a record company saying that. There would be many, many genres of music that would have to be removed, whether it was country singers singing about conservative politics, or whether it was rap artists rapping about life in an urban environment."

13. Clark, "Core Values of the Casual Industry," 2006.

14. Richards, "Profile|J. K. Rowling," 2000.

Appendix A: Player Survey

1. Mittell, *Genre and Television*, 2004, 94–95.

2. There also is no data for how Gamezebo players compare to players on other casual game websites.

3. This result technically can be the result of game developers targeting the stereotype of casual players and therefore erring on the side of making games too easy for the audience.

References

Aarseth, Espen. "Playing Research: Methodological Approaches to Game Analysis." Presented at the Digital Arts & Culture conference, Melbourne, 2003. http://hypertext.rmit.edu.au/dac/papers/Aarseth.pdf.

Aarseth, Espen, Solveig Marie Smedstad, and Lise Sunnanå. "A Multidimensional Typology of Games." In *Level Up Conference Proceedings*, 48–53. Utrecht: Utrecht University, 2003.

Addison, Amy. "Confessions of a Non-Gaming Mom." *Game Career Guide*, August 5, 2008. http://www.gamecareerguide.com/features/585/confessions_of_a_nongaming_.php.

Apperley, Thomas, Catherine Beavis, Clare Bradford, Joanne O'Mara, and Christopher Walsh. "Researching Kids and Videogames: Games, Game Play and Literacy in the Twenty First Century." In *Proceedings of the [player] Conference*, 4–27. Copenhagen: IT University of Copenhagen, 2008.

area/code. *Parking Wars*. A&E Television Networks (Facebook), 2008. http://www.new.facebook.com/apps/application.php?id=31435010008.

Atari. *Centipede*. Arcade. Atary, 1980.

Atari. *Pong*. Arcade. Atari, 1972.

Barr, Alfred H. *Cubism and Abstract Art*. New York: Arno Press, 1966.

Beaudouin-Lafon, Michel. "Instrumental Interaction: An Interaction Model for Designing Post-WIMP User Interfaces." In *Proceedings of the SIGCHI Conference on Human Factors in Computing Systems*, 446–453. The Hague, The Netherlands: ACM, 2000.

Bell, Erin. "Jewel Quest 2 Review." *GameZebo*, June 12, 2008. http://www.gamezebo.com/games/jewel-quest-2/review.

Big Fish Games. "Big Fish Games Website," October 11, 2008. http://www.bigfishgames.com.

Big Fish Games. "Puzzle Games|Big Fish Games," 2008. http://www.bigfishgames
.com/download-games/genres/4/puzzle.html.

Big Fish Studios. *Atlantis*. Windows. 2005.

Big Fish Studios. *Azada*. Windows. Big Fish Games, 2007.

Big Fish Studios. *Azada: Ancient Magic*. Windows. Big Fish Games, 2008.

Big Fish Studios. *Mystery Case Files: Huntsville*. Windows. Big Fish Games, 2005.

Big Splash Games. *Chocolatier*. Windows. PlayFirst, Inc., 2007.

Blizzard Entertainment. *WarCraft III*. Windows. Blizzard Entertainment, 2002.

Blizzard Entertainment. *World of Warcraft*. Windows. Blizzard Entertainment, 2004.

Boyer, Brandon. "NPD: 72% Of of U.S. Plays Games, Only 2–3% Own Multiple Consoles." *Gamasutra*, April 2, 2008. http://www.gamasutra.com/php-bin/news_index.php?story=18107.

Braben, David, and Ian Bell. *Elite*. Commodore 64. Firebird, 1985.

Bulletproof Software. *Yoshi's Cookie*. Nintendo Entertainment System. Nintendo, 1992.

Burrill, Derek A.. *Die Tryin': Videogames, Masculinity, and Culture*. New York: Peter Lang Publishing, 2008.

Cadogan, Adelaide. *Illustrated Games of Patience*, 3rd ed. London: Sampson Low, 1876.

Cadogan, Adelaide. *Lady Cadogan's Illustrated Games of Solitaire or Patience: New Rev. Ed., Including American Games*. Philadelphia: D. McKay, 1914.

Capcom. *Super Puzzle Fighter II Turbo*. Arcade. Capcom, 1996.

Carbonated Games. *Hexic*. Windows. Microsoft Game Studios, 2003.

Carless, Simon. "In-Depth: Inside Puzzle Quest—The Postmortem." *GameSet-Watch*, March 26, 2008. http://www.gamesetwatch.com/2008/03/in_depth_inside_puzzle_quest_the_postmortem.php.

Casamassina, Matt. "Wii Sports Review. Simple. Fun. But Does It Have Any Depth?" *ign*, November 13, 2006. http://wii.ign.com/articles/745/745708p1.html.

Cassell, Justine, and Henry Jenkins, eds. *From Barbie to Mortal Kombat: Gender and Computer Games*. Cambridge, MA: MIT Press, 1998.

Cerny, M., and M. John. "Game Development: Myth vs. Method." *Game Developer* (June 2002): 32–36.

Cifaldi, Frank. "Popping in on PopCap: James Gwertzman on Casual Growth." *Gamasutra*, December 14, 2005. http://www.gamasutra.com/features/20051214/cifaldi_pfv.htm.

Clark, Ethan. "Core Values of the Casual Industry: Q&A Download with Scott Bilas." *Casual Connect Magazine* (Fall 2006). http://mag.casualconnect.org/fall2006/Fall_2006_Kiev_english.pdf.

Codeminion. *Magic Match*. Windows. Microsoft, 2005.

Compile. *Puyo Puyo*. MSX. 1991.

Consalvo, Mia. *Cheating: Gaining Advantage in Videogames*. Cambridge, MA: MIT Press, 2007.

Costikyan, Greg. "Burning Down the House—Game Developers Rant." Presented at the Game Developers Conference, San José, March 11, 2005. http://crystaltips.typepad.com/wonderland/2005/03/burn_the_house_.html.

Costikyan, Greg. "Game Styles, Innovation, and New Audiences: An Historical View." In *Proceedings of DiGRA 2005 Conference: Changing Views—Worlds in Play*. Vancouver, 2005. http://www.digra.org/dl/db/06278.11155.pdf.

Culin, Stewart. "Mancala: The National Game of Africa." In *The Study of Games*, ed. Elliott M Avedon and Brian Sutton-Smith, 94–108. New York: J. Wiley, 1971.

Cyan. *Myst*. Macintosh. Brøderbund, 1993.

Davis, J. P., K. Steury, and R. Pagulayan. "A Survey Method for Assessing Perceptions of a Game: The Consumer Playtest in Game Design." *Game Studies: The International Journal of Computer Game Research* 5, no. 1 (2005). http://www.gamestudies.org/0501/davis_steury_pagulayan/.

Dill, Karen E. "Violent Video Games Can Increase Aggression." American Psychological Association, 2000. http://www.apa.org/releases/videogames.html.

Dillon, Beth A. "Casuality: Luxor, Mah Jong Quest, Fish Tycoon Devs Talk Postmortems." *Gamasutra*, June 29, 2006. http://www.gamasutra.com/php-bin/news_index.php?story=9913.

DMA Design. *Grand Theft Auto III*. PlayStation 2. Rockstar Games, 2001.

Dobson, Jason. "Study: 'Casual' Players Exhibit Heavy Game Usage." *Gamasutra*, June 28, 2006. http://www.gamasutra.com/php-bin/news_index.php?story=9893.

Ducheneaut, Nicolas, Nick Yee, Eric Nickell, and Robert J. Moore. "Building an MMO with Mass Appeal: A Look at Gameplay in World of Warcraft." *Games and Culture* 1, no. 4 (October 1, 2006): 281–317.

Eco, Umberto. *The Search for the Perfect Language (The Making of Europe)*. Oxford: Blackwell, 1995.

Eliot, T. S. "Philip Massinger." In *In the Sacred Wood: Essays on Poetry and Criticism*, 125. London: Methune, 1920.

Electronic Arts Los Angeles. *Boom Blox*. Wii. Electronic Arts, 2008.

ELIZABETH "LIZZIE." "FUN? WELL.... DIFFERENT.... THAT'S FOR SURE !!," 2008. http://www.amazon.com/review/product/B000GH3PYA/ref=cm_cr_dp_all _helpful?%5Fencoding=UTF8&coliid=&showViewpoints=1&colid=&sortBy =bySubmissionDateDescending.

eMarketer. *Importance of Select Factors to US Console Video Gamers When Purchasing Next-Generation Video Game Console, March–April 2008 (% of respondents)*, June 20, 2008. http://www.emarketer.com/Chart.aspx?id=77015.

Enevold, Jessica, and Charlotte Hagström. "My Momma Shoots Better than You! Who Is the Female Gamer?" In *Proceedings of the [Player] Conference*, 144–167. Copenhagen: IT University of Copenhagen, 2008.

Entertainment Software Association. *Essential Facts about the Computer and Video Game Industry*, 2008. http://www.theesa.com/facts/pdfs/ESA_EF_2008.pdf.

Epic Games. *Gears of War*. Xbox. Microsoft Game Studios, 2006.

Exidy. *Destruction Derby*. Arcade. Exidy, 1975.

Flaherty, Jessica. "Flaherty: Stay Cool: Come to the Senior Center!—Medford, MA—Medford Transcript." *Wicked Local Medford*, June 25, 2008. http://www .wickedlocal.com/medford/archive/x222995948/Flaherty-Stay-cool-Come-to-the -Senior-Center.

Flanagan, Mary. "Troubling 'Games for Girls': Notes from the Edge of Game Design." In *Proceedings of DiGRA 2005 Conference: Changing Views—Worlds in Play*. Vancouver, 2005. http://www.digra.org/dl/db/06278.14520.pdf.

Friends Games. *Magic Match: The Genie's Journey*. Windows. Oberon Games, 2007.

Fron, Janine, Tracy Fullerton, Jacquelyn Ford Morie, and Celia Pearce. "The Hegemony of Play." In *Situated Play: Proceedings of the Third International Conference of the Digital Games Research Association*, ed. Akira Baba, 309–318. Tokyo, 2007. http://www.digra.org/dl/db/07312.31224.pdf.

Fullerton, Tracy, Chris Swain, and Steven Hoffman. *Game Design Workshop: Designing, Prototyping, and Playtesting Games*. San Francisco: CMP Books, 2004.

Gabler, Kyle, Kyle Gray, Matt Kucic, and Shalin Shodhan. "How to Prototype a Game in Under 7 Days." *Gamasutra*, October 26, 2005. http://www.gamasutra .com/features/20051026/gabler_pfv.htm.

GameDaily Staff. "Are Big Budget Console Games Sustainable?" *GameDaily*, March 10, 2006. http://biz.gamedaily.com/industry/advertorial/?id=12089.

Gamehouse. *Collapse*. Windows. 1998.

Gamelab. *Diner Dash*. Windows. PlayFirst, Inc., 2003.

Gamelab. *Shopmania*. Windows. iWin, 2006.

Garfield, Richard. "Metagames." In *Horsemen of the Apocalypse: Essays on Roleplaying*, ed. by Jim Dietz, 14–21. Charleston, IL: Jolly Rogers Games, 2000. http://www.gamasutra.com/features/gdcarchive/2000/garfield.doc.

Gartenberg, Michael. *Simple Games: The Return to Fun*. Jupiter Research, March 15, 2007. http://www.jupiterresearch.com/bin/item.pl/research:concept/111/id =98905.

Geryk, Bruce. "A History of Real-Time Strategy Games." *GameSpot*, 2001. http://www.gamespot.com/gamespot/features/all/real_time/.

Gibson, Ellie. "Phil Harrison on the Future of PlayStation Interview." *Eurogamer*, December 14, 2005. http://www.eurogamer.net/article.php?article_id=62158.

Goble, Brian, and Garrett Price. "Build-a-Lot Post Mortem." *Casual Games Quarterly* 3, no. 1 (2008): 2–5. http://www.igda.org/casual/quarterly/3_1/igda_casual _game_quarterly_3_1.pdf.

Goldstein, Patrick. "Why Is Pac-Man Grinning? He's Sharing His Quarters." *Los Angeles Times*, February 4, 1982, H1.

Harmetz, Aljean. "Is Electronic-Games Boom Hurting the Movies?" *New York Times*, July 6, 1981, C11.

Harmonix Music Systems, Inc. *Guitar Hero*. PlayStation 2. RedOctane Inc., 2005.

Harmonix Music Systems, Inc. *Rock Band*. Xbox 360. MTV Games, 2007.

Harmonix Music Systems, Inc. *Rock Band 2*. Xbox 360. MTV Games, 2008.

Hatfield, Daemon. "Puzzle Quest: Challenge of the Warlords Review." *ign.com*, March 29, 2007. http://ds.ign.com/articles/777/777079p1.html.

Herdlick, Catherine, and Eric Zimmerman. "Redesigning Shopmania: A Design Process Case Study." *IGDA Casual Games Quarterly* 2, no. 1 (2006). http://www .igda.org/casual/quarterly/2_1/index.php?id=6.

HipSoft. *Build-a-Lot*. Windows. Big Fish Games, 2007.

Hot Lava Games. *7 Wonders of the Ancient World*. Windows. 2006.

Howson, Greg. "Casual Games Rule Charts." *Guardian Games*, July 31, 2008. http://blogs.guardian.co.uk/games/archives/2008/07/31/casual_games_rule_charts .html.

IGDA Casual Games SIG. "2005 Casual Games White Paper," 2005. http://www.igda.org/casual/IGDA_CasualGames_Whitepaper_2005.pdf.

Infinite Interactive. *Puzzle Quest*. Windows. D3 Publisher, 2007.

Intelligent Systems. *Panel de Pon*. Super Nintendo Entertainment System. Nintendo, 1995.

Itzkoff, Dave. "Rec-Room Wizard." *The New York Times*, August 10, 2008, sec.Arts/Television. http://www.nytimes.com/2008/08/10/arts/television/10itzk.html?pagewanted=all.

iWin. *Jewel Quest*. Windows. Gamehouse, 2004.

iWin. *Jewel Quest II*. Windows. iWin, 2007.

Jakobsson, Mikael. "Playing with the Rules: Social and Cultural Aspects of Game Rules in a Console Game Club." In *Situated Play: Proceedings of the Third International Conference of the Digital Games Research Association*, ed. Akira Baba, 386–392. Tokyo, 2007. http://www.digra.org/dl/db/07311.01363.pdf.

Järvinen, Aki. "Communities of Nurturing: How to Design Empathy." Presented at the Nordic Game Conference, Malmö, 2007.

Järvinen, Aki. "Games without Frontiers: Theories and Methods of Game Studies and Game Design." PhD dissertation, University of Tampere, 2008. http://acta.uta.fi/english/teos.php?id=11046.

Jenkins, Henry. "Complete Freedom of Movement: Video Games as Gendered Play Spaces." In *From Barbie to Mortal Kombat: Gender and Computer Games*, ed. Justine Cassell and Henry Jenkins, 262–297. Cambridge, MA: MIT Press, 1998.

Jones, Whitmore. *Games of Patience for One or More Players*. Second Series. London: L. Upcott Gill, 1898.

Juul, Jesper. "A Certain Level of Abstraction." In *Situated Play: Proceedings of the Third International Conference of the Digital Games Research Association*, ed. Baba Akira, 510–515. Tokyo, 2007. http://www.jesperjuul.net/text/acertainlevel/.

Juul, Jesper. "Fear of Failing? The Many Meanings of Difficulty in Video Games." In *The Video Game Theory Reader 2*, ed. Bernard Perron and Mark J. P. Wolf, 237–252. New York, NY: Routledge, 2008.

Juul, Jesper. *Half-Real: Video Games between Real Rules and Fictional Worlds*. Cambridge, MA: MIT Press, 2005.

Juul, Jesper. "The Open and the Closed: Games of Emergence and Games of Progression." In *Computer Game and Digital Cultures Conference Proceedings*, ed. Frans Mäyrä, 323–329. Tampere: Tampere University Press, 2002.

Juul, Jesper, and Marleigh Norton. "Easy to Use and Incredibly Difficult: On the Mythical Border between Interface and Gameplay," Presented at the Foundations of Digital Games Conference, Port Canaveral, FL, 2009.

Kafai, Y. B., C. Heeter, J. Denner, and J. Y. Sun. *Beyond Barbie and Mortal Kombat: New Perspectives on Gender and Computer Games.* Cambridge, MA: MIT Press, 2008.

Kallio, Kirsi Pauliina, Kirsikka Kaipainen, and Frans Mäyrä. *Gaming Nation? Piloting the International Study of Games Cultures in Finland.* Hypermedia Laboratory Net Series, 2007.

Kapalka, Jason. "10 Ways to Make a BAD Casual Game." *Casual Connect Magazine*, Summer 2006. http://www.casualconnect.org/content/gamedesign/kapalka-tenways.html.

Kim, Scott. "Games for the Rest of Us: Puzzles, Board Games, Game Shows." Presented at the Computer Game Developers Conference, Long Beach, CA, 1998. http://www.scottkim.com/thinkinggames/cgdc98.html.

Konami. *Dance Dance Revolution.* Arcade. Konami Digital Entertainment, 1998.

Konami. *Scramble.* Arcade. Konami, 1981.

Konami. "Scramble Installation Manual," 1981. http://www.arcadedocs.com/vidmanuals/S/Scramble.pdf.

Kotaku. "Kotaku, the Gamer's Guide," January 1, 2007. http://web.archive.org/web/20070101172621/http://kotaku.com/.

Kuchera, Ben. "WiiWare and VC Releases (6-30-08 Furious Magnetic Edition)." *Ars Technica: Opposable Thumbs*, June 30, 2008. http://arstechnica.com/journals/thumbs.ars/2008/06/30/wiiware-and-vc-releases-6-30-08-furious-magnetic-edition.

Kuittinen, Jussi, Annakaisa Kultima, Johannes Niemelä, and Janne Paavilainen. "Casual Games Discussion." In *Proceedings of the 2007 Conference on Future Play*, 105–112. Toronto, Canada: ACM, 2007.

Lady Cadogan. *Lady Cadogan's Illustrated Games of Solitaire or Patience: New Rev. Ed., including American Games.* Philadelphia: D. McKay, 1914.

Lafferty, Michael. "World of Warcraft Review—PC." *GameZone*, December 13, 2004. http://pc.gamezone.com/gzreviews/r19235.htm.

Lane, R. D., P. M. Chua, and R. J. Dolan. "Common Effects of Emotional Valence, Arousal and Attention on Neural Activation during Visual Processing of Pictures." *Neuropsychologia* 37, no. 9 (August 1999): 989–997.

Lazzaro, Nicole. "Why We Play Games: Four Keys to More Emotion in Player Experiences." Presented at the Game Developers Conference, San José, CA, 2004. http://www.xeodesign.com/xeodesign_whyweplaygames.pdf.

Lenhart, A., J. Kahne, E. Middaugh, A. Macgill, C. Evans, and J. Vitak. "Teens, Video Games, and Civics." *Pew Internet and American Life Report* (2008). http://www.pewinternet.org/pdfs/PIP_Teens_Games_and_Civics_Report_FINAL.pdf.

Lohr, Steve. "Computer Games Venture into the World of Gun, Bow and Big Game." *New York Times*, March 29, 1999, C1.

Los Angeles Times. "New Video Games Sweep Nation." *Los Angeles Times*, November 28, 1974, CS_A9.

Malaby, Thomas. "Beyond Play: A New Approach to Games." *Games and Culture* 2, no. 2 (2007): 95.

Marriott, Michel. "The Un-Doom Boom." *New York Times*, June 25, 2003, G1.

Maxis. *SimCity*. DOS. Brøderbund, 1989.

Maxis. *The Sims 2*. Windows. Electronic Arts, 2004.

McElroy, Justin. "Behind the Game: Bejeweled®." *GameZebo*, September 21, 2006. http://www.gamezebo.com/features/behind_the_game_bejeweled.html.

Meretzky, Steve. "What Is a Casual Game?" Presented at the Game Developers Conference, San Francisco, 2007.

Microprose Software. *Pirates!* Commodore 64. Microprose Software, 1987.

Microsoft. "Microsoft Reveals First Details of Next-Generation Xbox: Company's Chief XNA Architect Shares Vision for HD Era of Gaming," March 9, 2005. http://www.microsoft.com/presspass/press/2005/mar05/03-09GDC05PR.mspx.

Microsoft. *Solitaire*. Windows. Microsoft, 1990.

Midway. *Ms. Pac-Man*. Arcade. Midway, 1981.

Midway. *Sea Wolf*. Arcade. Midway, 1977.

Mirabella III, Fran. "IGN: Dr. Mario 64 Review." *ign.com*, April 17, 2001. http://ign64.ign.com/articles/165/165952p1.html.

Mitchell. *Puzz Loop*. Arcade. Mitchell, 1998.

Mittell, Jason. *Genre and Television*. New York: Routledge, 2004.

MobyGames. "Falling Block Puzzles," 2008. http://www.mobygames.com/game-group/falling-block-puzzles.

MobyGames. "Puzz Loop Variants," 2008. http://www.mobygames.com/game-group/puzz-loop-variants.

MobyGames. "Tile-Matching Puzzle Games (Creation)," 2008. http://www.mobygames.com/game-group/tile-matching-puzzle-games-creation.

Moribe, Kuniaki. *Chain Shot!* PC-98. 1985.

Mumbo Jumbo. *Luxor*. Windows. GameHouse, Inc., 2005.

Nakamura, Rika, and Hanna Wirman. "Girlish Counter-Playing Tactics." *Game Studies* 5, no. 1 (2005). http://www.gamestudies.org/0501/nakamura_wirman/.

Namco. *Donkey Konga.* GameCube. Nintendo, 2003.

Namco. *Pac-Man.* Arcade. Namco, 1980.

Neumann, John Von, and Oskar Morgenstern. *Theory of Games and Economic Behavior.* Princeton: Princeton University Press, 1944.

Neversoft. *Guitar Hero III.* Xbox 360. Activision, 2007.

Nielsen Games. *Video Games in Europe 2008.* Interactive Software Federation of Europe, 2008. http://www.isfe-eu.org/index.php?oidit=T001:662b16536388a726092159932136591I.

Nielsen, Jakob. *Usability Engineering.* San Diego, CA: Academic Press, Inc., 1993.

Nintendo. *Dr. Mario.* Nintendo Entertainment System. Nintendo, 1990.

Nintendo. *Financial Results Briefing for the Nine-Month Period Ended December 2007. Supplementary Information,* January 25, 2008. http://www.nintendo.co.jp/ir/pdf/2008/080125e.pdf.

Nintendo. *Wii Sports.* Wii. Nintendo, 2006.

Nintendo EAD. *Animal Crossing.* GameCube. Nintendo, 2002.

Nintendo EAD. *Super Mario 64.* Nintendo 64. Nintendo, 1996.

Nintendo EAD. *Wii Fit.* Wii. Nintendo, 2008.

Nintendo EAD. *Wii Play.* Wii. Nintendo, 2006.

Nintendo EAD Tokyo. *Super Mario Galaxy.* Wii. Nintendo, 2007.

Norman, Donald A. *Emotional Design: Why We Love (or Hate) Everyday Things.* New York: Basic Books, 2005.

Norman, Donald A. *The Design of Everyday Things.* New York: Basic Books, 2002.

Oberon Media. "GAME DEVELOPERS SUBMISSION GUIDELINES," 2007. http://www.oberongames.com/publishing.asp.

Olsen, Neil. "Anti-Peggleite." *PC Game Magazine,* April 8, 2008.

Onesound. "Animal Crossing is Tragic," 2005.

Orbanes, Philip E. *The Game Makers: The Story of Parker Brothers, from Tiddledy Winks to Trivial Pursuit.* Boston: Harvard Business School Press, 2003.

Pajitnov, Alexey, and Vadim Gerasimov. DOS. *Tetris,* 1985.

Parker Brothers. *Monopoly.* Board game. Parker Brothers, 1936.

Parlett, David. *A History of Card Games*. Oxford: Oxford University Press, 1990.

Pearce, Celia. "The Truth about Baby Boomer Gamers: A Study of Over-Forty Computer Game Players." *Games and Culture* 3, no. 2 (April 1, 2008): 142–174.

PF Magic. *Catz: Your Computer Petz*. Windows 3.1. Virgin Interactive Entertainment, 1996.

PF Magic. *Dogz: Your Computer Pet*. Windows 3.1. Virgin Interactive Entertainment, 1995.

Piccione, Peter A. "In Search of the Meaning of Senet." *Archaeology* 33 (July/August 1980): 5558.

PopCap Games. *Bejeweled*. Windows. 2001.

PopCap Games. *Bejeweled 2 Deluxe*. Windows. 2004.

PopCap Games. *Peggle*. Windows. 2007.

PopCap Games. "PopCap's Bejeweled® Franchise Hits 25 Million Units Sold Mark." August 19, 2008. http://www.popcap.com/press/release.php?pid=428.

PopCap Games. "Publishing FAQ," 2006. http://developer.popcap.com/forums/pop_info.php#publishing.

PopCap Games. "Survey: Tens of Millions of 'White Collar' Workers Play 'Casual' Video Games—One in Four Play at Work, and Senior Execs Play Even More," September 4, 2007. http://www.popcap.com/press/release.php?pid=227.

PopCap Games. *Zuma*. Windows. 2004.

Pratchett, Rhianna. *Games in the UK*. BBC Audience Research, 2005. http://open.bbc.co.uk/newmediaresearch/files/BBC_UK_Games_Research_2005.pdf.

Preece, Paul. *Desktop Tower Defense*. Flash. 2007. http://www.handdrawngames.com/DesktopTD/Game.asp.

Psygnosis. *Wipeout*. PlayStation. Psygnosis, 1995.

Q Entertainment. *Lumines*. Sony PSP. Ubisoft, 2004.

Q Entertainment. *Lumines Live!* Xbox 360. Q Entertainment, 2006.

Q Entertainment. *Meteos*. Nintendo DS. Ubisoft, 2005.

Range, Peter Ross. "The Space Age Pinball Machine." *New York Times*, September 15, 1974.

Raptisoft. *Chuzzle Deluxe*. Windows. PopCap Games, 2005.

RCM. *Magic Jewelry*. NES. RCM, 1990.

Reflexive Entertainment. *Big Kahuna Reef.* Windows. 2004.

Relentless Software Ltd. *Buzz! The Mega Quiz.* PlayStation 2. Sony Computer Entertainment Europe, 2007.

Relentless Software Ltd. *Buzz!: Quiz TV.* PlayStation 3. Sony Computer Entertainment Europe, 2008.

Richards, Linda L. "Profile|J. K. Rowling." *January Magazine* (October 2000). http://januarymagazine.com/profiles/jkrowling.html.

Rockstar Games North. *Grand Theft Auto: San Andreas.* Windows. Take-Two Interactive, 2005.

Ryan, Kim. "Game Designers Focus on Girls." *San Francisco Chronicle,* October 5, 2008. http://www.sfgate.com/cgi-bin/article.cgi?f=/c/a/2008/10/05/BUMD13B33S.DTL.

Salen, Katie, and Eric Zimmerman. *Rules of Play: Game Design Fundamentals.* Cambridge, MA: MIT Press, 2004.

Sandlot Games. *Cake Mania 3.* Windows. Sandlot Games, 2008

Schell, Jesse. "Hardcore Games for Casual Audiences." Presented at the Austin Game Developers Conference, Austin, TX, 2008. http://www.gamedev.net/reference/business/features/08AGDC1/page6.asp.

Scott, Leslie. *Jenga.* Milton Bradley, 1986.

Sega. *Baku Baku Animal.* Arcade. Sega, 1995.

Seydoux, Chaz. "shokkingu hitofude." *insert credit,* January 16, 2006. http://www.insertcredit.com/features/hitofude/.

Shneiderman, B. "Direct Manipulation: A Step beyond Programming Languages." *Computer* 16, no. 8 (1983): 57–69.

Sicart, Miguel. "Defining Game Mechanics". Game Studies: The International Journal of Computer Game Research 8, no. 21 (2008). http://gamestudies.org/0802/articles/sicart.

Singline, Karl, and Jason Hill. "Your Turn: A Casual Rant." *The Sydney Morning Herald Blogs: Screen Play,* September 19, 2008. http://blogs.smh.com.au/screenplay/archives//020098.html.

sinofsky. "If You Love RPGs and Puzzle Games You Will Love This Game!" *pricegrabber.com,* June 11, 2007. http://reviews.pricegrabber.com/nintendo-ds-games/m/34169778/.

Smith, James C. "CasualCharts.com." *CasualCharts,* 2008. http://www.game-sales-charts.com.

Smith, Jonas Heide. "Plans and Purposes: How Video Games Shape Player Behavior." PhD dissertation, IT University of Copenhagen, 2006.

Smith, Matthew. *Manic Miner*. Commodore 64. Bug-Byte, 1983.

Sonic Team. *Samba de Amigo*. DreamCast. SEGA Corporation, 1999.

Sony Computer Entertainment Inc. "SONY COMPUTER ENTERTAINMENT INC. TO LAUNCH ITS NEXT GENERATION COMPUTER ENTERTAINMENT SYSTEM, PLAYSTATION 3 IN SPRING 2006," May 16, 2005. http://www.us .playstation.com/News/PressReleases/279.

Sony Online Entertainment. *EverQuest*. Windows. Sony Online Entertainment, 1999.

Sony Pictures Digital. *The Da Vinci Code*. Windows. Sony Pictures Digital, 2006.

Staiger, Janet. *Perverse Spectators: The Practices of Film Reception*. New York: New York University Press, 2000.

Steinberg, Scott. "Puzzle Quest: Challenge of the Warlords Review." *Gamezebo*, November 7, 2007. http://www.gamezebo.com/reviews/puzzle_quest_challenge _of_the.html.

Steinmeyer, Phil. "Playing It Safe (Enough)." *Philsteinmeyer*, November 21, 2005. http://www.philsteinmeyer.com/17/playing-it-safe-enough/.

Success. *Zoo Keeper*. Game Boy Advance. Ignition Entertainment, 2003.

Sumo Digital. *Virtua Tennis 3*. Xbox 360. SEGA, 2007.

Suzie Q. "Jewel Quest 2 Reviews," 2007. http://www.pixelparadox.com/arcade _games/jewel_quest_2.htm.

Švelch, Jaroslav. "YOU BURNOUTS ROTTING IN FRONT OF YOUR COMPUTERS!: Cult of the Superhardcore Gamer in Czech Gaming Culture." *Different Gaming*, January 21, 2008. http://differentgaming.blogspot.com/2008/01/you -burnouts-rotting-in-front-of-your.html.

Taito. *Plotting*. Arcade. Taito, 1989.

Taito. *Puzzle Bobble*. Arcade. Taito, 1994.

Taito. *Puzznic*. Arcade. Taito, 1989.

Tarbart. *Games of Patience*. London: T. De La Rue, 1901.

Taylor, T. L. *Play Between Worlds: Exploring Online Game Culture*. Cambridge, MA: MIT Press, 2006.

The WizardWorks Group, Inc. *Deer Hunter*. DOS. The WizardWorks Group, Inc., 1997.

Three Rings Design. *Puzzle Pirates*. Java. Three Rings Design, 2003.

Totilo, Stephen, and Matt Elias. "Slash, Developers Riff on 'Guitar Hero III': GameFile." *MTV News*, July 31, 2007. http://www.mtv.com/news/articles/1565980/20070731/velvet_revolver.jhtml.

Tufte, Edward R. *Beautiful Evidence*. Cheshire, CT: Graphics Press, 2006.

Tunnell, Jeff. "Five Foundational Steps to Surviving as a Game Developer." *Make It Big in Games*, February 7, 2006. http://www.makeitbigingames.com/blog/?p=14.

Ubisoft Montpellier. *Rayman Raving Rabbids*. Wii. Ubisoft, 2006.

USA Today. "Nintendo Hopes Wii Spells Wiinner." *USA Today*, August 15, 2006. http://www.usatoday.com/tech/gaming/2006-08-14-nintendo-qa_x.htm.

Valve. *Half-Life 2: Episode One*. Windows. Valve Corporation, 2006.

Valve. "Half-Life 2: Episode One Stats," 2008. http://www.steampowered.com/status/ep1/.

Varney, Allen. "Attack of the Parasites." *The Escapist*, February 28, 2006. http://www.escapistmagazine.com/articles/view/issues/issue_34/207-Attack-of-the-Parasites.

VGChartz.com. "VGChartz.com," February 24, 2009. http://www.vgchartz.com/.

Wildfire Studios. *Tumblebugs*. Windows. GameHouse, Inc., 2005.

Young, Bryan-Mitchell. "The Disappearance and Reappearance and Disappearance of the Player in Videogame Advertising." In *Situated Play: Proceedings of the Third International Conference of the Digital Games Research Association*, ed. Akira Baba, 235–242. Tokyo, 2007. http://www.digra.org/dl/db/07312.01482.pdf.

Zelfden, N. Evan Van. "Gingold Talks Spore's 'Magic Crayon' Approach." *Gamasutra*, June 26, 2008. http://www.gamasutra.com/php-bin/news_index.php?story=19122.

Index

and the target audience, 10–11, 74–
76, 151, 177, 182–183, 193, 194–195,
199–200, 202, 204–205
Game Developers Conference, 13, 25,
184–185
Game development budgets, 148
Gamehouse, 185
Game mechanics, 68, 72, 79, 84, 98,
100–101, 126, 150
Games
casual (*see* Casual games)
defined, 131–132
design time of, 74–78
difficulty of (*see* Difficulty and
punishment)
fitting into players' lives, 5, 10, 72
flexibility of (*see* flexibility of games
and players)
game-playing time, 77
goals, 23
hardcore (*see* Hardcore games)
historical time of, 77
as languages, 138–139
meaning of, 121–128
as mental workout, 163, 167, 174
open box, 79, 110–113, 203
and players, 9, 52–55, 78, 146–147
Game theory (economic), 53n72
Gamezebo, 65, 153
Garfield, Richard, 121–122
Gears of War, 31–33, 49, 117
Gender. *See* Players
Genre, 33, 65–68, 78, 79, 84–85, 98–
100
Gingold, Chaim, 113–116
Go (game), 197
Goal orientation (when playing), 126–
127
Goals, 129–143, 133n11
presentation of, 133–138
problem with, 138–139
shared understanding of, 126
Graphics, 12–16, 26, 148, 211
high definition, 13–16
and innovation, 14
three-dimensional, 16

Grand Theft Auto series, 23, 130–131,
134–136, 138–139, 176, 208
Guitar Hero series, 5, 20, 22, 23, 37,
45, 56, 59, 79, 103–107, 110–118,
129–130, 145, 158, 168, 179–182,
195, 207–208, 214
multiple ways to play, 139–143

Hagström, Charlotte, 10n17
Half-Life 2: Episode One, 148, 179
Halo, 169
Hardcore games, 8–10, 53–55, 103,
190
affordances of, 53–55
conditions for developing, 7, 151,
178–179, 205
inflexibility of, 10, 53–55, 130
played casually, 139–143
Hardcore players, 8–10, 28–30, 51–55
ethic of, 28–29, 62, 143, 180
flexibility of, 10, 53–55
lapsed, 12, 51–52, 157, 162–163, 176,
215
stereotype of, 8–10, 28–29, 146
Harmonix, 116, 142–143, 179–182, 207
Harry Potter, 151
Herdlick, Catherine, 39
Hidden object games, 1, 68, 79, 100,
172, 212–213
Hoffman, Steven, 53n72

Innovation and cloning, 14, 67, 84,
92–97, 172, 206, 212
Insaniquarium, 165
International Game Developers
Association, 25n2, 92n31, 184
Interruptibility, 30, 36–39, 50, 57–58.
See also Time commitment
iWin, 92

Jakobsson, Mikael, 53n71
Järvinen, Aki, 68n10, 124
Jenga, 33
Jenkins, Henry, 10n17
Jewel Quest series, 40, 92
John, Michael, 75